STATISTICAL MECHANICS FOR BEGINNERS

A Textbook for Undergraduates

STATISTICAL MECHANICS FOR BEGINNERS

A Textbook for Undergraduates

Lucien Gilles Benguigui

Israel Institute of Technology, Israel

World Scientific

NEW JERSEY · LONDON · SINGAPORE · BEIJING · SHANGHAI · HONG KONG · TAIPEI · CHENNAI

Published by

World Scientific Publishing Co. Pte. Ltd.

5 Toh Tuck Link, Singapore 596224

USA office: 27 Warren Street, Suite 401-402, Hackensack, NJ 07601

UK office: 57 Shelton Street, Covent Garden, London WC2H 9HE

British Library Cataloguing-in-Publication Data
A catalogue record for this book is available from the British Library.

ISBN-13 978-981-4299-11-4
ISBN-10 981-4299-11-1
ISBN-13 978-981-4299-12-1 (pbk)
ISBN-10 981-4299-12-X (pbk)

Typeset by Stallion Press
Email: enquiries@stallionpress.com

Printed in Singapore.

To Lucien Godefroy who introduced me to Statistical Mechanics.

Preface

This book is intended for students who begin for the first time the study of statistical mechanics. There are two different approaches to teach thermal physics to the physics students at the level of BSc. The first is to expose the subject in two separate courses: one in thermodynamics, i.e. the macroscopic aspect of thermal physics, and a second one (taught immediately after the first) in statistical mechanics, i.e. the microscopic aspect. There are excellent books that choose this way. The other approach is to present the subject in only one course, mixing the two aspects. Excellent books that follow this more compact method also exist. Here there is no place to discuss the advantages and the disadvantages of each of the two approaches. In this book, I follow the first one. This means that this course is suited for students that already have some knowledge in thermodynamics.

Historically, classical statistical mechanics was first developed and only later, with the progress of quantum theory, the quantum statistical mechanics was born. I think that, from a pedagogical point of view, it is easier to teach the quantum statistical mechanics than its classical counterpart. This is the reason why the main part of this book is devoted to the quantum statistical mechanics. It suffices that the student has elementary knowledge of the basic results of the quantum theory to be able to understand the matter.

This book may appear very short. In fact, it is effectively far form being complete. For example, in thermodynamics the concept of entropy is introduced in connection with irreversible process.

However, in this book, I did not discuss this problem, giving time for student to study it later.

I tried to present the subject in a consistent form: first the general principles or the methods giving the links between the macroscopic and the microscopic worlds. In addition, in the second part, applications to simple situations are developed. It is good to give first the foundations and only afterwards the details of the applications. On the other side, I present classical cases as particular situations of quantum cases. This is not the way in which the matter is frequently taught. I think that the actual presentation has some novel aspect. The mathematical level is not very high. The reader has to be familiar with algebraic calculus, combinatorics, differential and integral calculus.

The book is almost exclusively for students. It is based on my personal teaching at the Technion. At the disposal of the teacher, there are many very good books with a lot of complementary details for an oral teaching in the classroom. But I did not find a book that I can recommend to the students when, for example, he was not able to assist some classes. When I taught this course, I had only two hours a week (and one hour for exercises) during one semester of 14 weeks. In such limited time, only the main points may be taught. This means that all is important in the book. It represents what a student needs to know in order to be able to follow others courses in his studies toward his first degree in physics (for example, a course in solid state physics). I introduced exercises which are straightforward applications of the matter of each chapter. They will help the student to assimilate the main concepts and methods.

I added a special chapter on the history of statistical mechanics. Since in the book itself I do not follow the historical development, I thought this could be interesting to bring about some views on how the theory was built.

I thank J. Unffick of the University of Utrecht for his help in preparing the chapter on the history of the statistical mechanics.

L. G. Benguigui

Haifa 2009

Contents

List of Figures

List of Figures

Physical Constants

Boltzmann constant	$k_B = 1.3896 \times 10^{-23}\,\text{J/K}$
	$k_B = 8.517 \times 10^{-5}\,\text{eV/K}$
Planck's constant	$h = 6.62607 \times 10^{-34}\,\text{J s}$
	$\hbar = h/2\pi = 1.0546 \times 10^{-34}\,\text{J s}$
Electron mass	$m = 9.1094 \times 10^{-31}\,\text{kg}$
Proton mass	$M = 1.6726 \times 10^{-27}\,\text{kg}$
Electric charge of electron	$e = 1.6021 \times 10^{-19}\,\text{C}$
Avogadro's number	$N_A = 6.022 \times 10^{23}$
Speed of light	$c = 2.9979 \times 10^{8}\,\text{m/s}$
Gas constant	$R = N_A\,k_B = 8.3145\,\text{J/K}$
Electron volt	$\text{eV} = 1.6021 \times 10^{-19}\,\text{J}$

Physical Constants

Introduction

In thermodynamics it is shown that the thermal properties of a system compound of a very large number of particles is characterized by a relatively small number of quantities such as the internal energy, the temperature, the entropy, the volume, the pressure *etc.* These are the macroscopic parameters of the matter. Thermodynamics was developed without any hypothesis about a microscopic picture of the matter in its three forms: solid, liquid or gas. However with the development of the atomic theory, it began to be possible to look for the link between the macroscopic world and a microscopic picture. At the end of the 19th and at the beginning of the 20th century, the first steps toward a theory relating the macroscopic world and a microscopic picture were proposed by Boltzmann and Gibbs. At this time the term *Statistical Mechanics* was coined by Gibbs.

The basic problem in statistical mechanics is to find the macroscopic properties of a system of particles from the knowledge of their microscopic properties. But at the microscopic level, the number of parameters is enormous. It is impossible, to follow each particle individually and to calculate the properties of the system by some average over all the particles. In front of this impossible task, it was necessary to proceed in another way. One has to leave the microscopic individual picture and to rely on a statistical approach.

The theory is based on a fundamental hypothesis. It is possible to formulate it as follows. The matter, whatever its state, is compounded of microscopic entities which have specific characteristics. The state of each entity does not remain the same but changes with

1

the time. In other words, the microscopic world is disordered. However it is admitted that the collective properties do remain stationary. For example, the molecules of a gas change constantly their velocities because of the collisions between them, but their mean velocity is well defined.

A particular theory which is based on some postulates was developed. The test of the validity of this method is the comparison between the consequences drawn from the postulates and the experimental results. The close agreement, which was found, is a guarantee of the validity of the method.

An important consequence of this approach is that thermal phenomena have their origin in mechanics. It is not an obvious thinking and the reader must be ready to adopt it. If a system is made of a relatively small number of bodies, one can apply the law of mechanics like in the planetary system. But if now the system is compounded of a very large number of units, one has the thermal phenomena. Some subtle points remain in this approach and in the past they were subjects of intense debate. In this book we shall not consider them except in the historical part. We think that in a first contact with this field, it is better to consider only the basic concepts. We hope that this first encounter with statistical mechanics will help the reader to be able to read more advanced books.

The book is divided into two parts: first the principles of the theory and in the second some applications. This does not correspond to the historical development of statistical mechanics as it is frequently presented. In order to follow the history, we added a chapter which presents those scientists who contributed to the main steps in the development of the theory.

The Thermodynamic Potentials.

Before embarking on the exposition of the theory and its applications, important results concerning the thermodynamic potentials are revised.

If the equilibrium state of the system is defined by the knowledge of some variables, there exists a function of these variables from which all the properties of the system can be deduced. These functions play the role of a potential for the following reason. If a perturbation appears in the system and the chosen variables are kept constant, the equilibrium state is reached when the potential is in a minimum. We shall consider three cases, which are important for the theory of statistical mechanics.

1. The entropy as a thermodynamic potential

In the first case, the state of the system is controlled by the extensive variables like the energy E, the volume V, the number of particles N, etc. (we took only these three variables by convenience). In such case, one considers the system as a closed system since no energy or no particle can enter or leave the system. The extensive variables are those which are proportional to the size of the system. The thermodynamic potential associated with such a state is the entropy $S(E, V, N)$. We recall that in a closed system, equilibrium is reached when S is maximum or when $-S$ is minimum. Entropy is a complex concept which may be presented in several ways. In the framework of this book, entropy is defined as a thermodynamic potential in the particular context of the closed system.

The entropy is a homogeneous function such that

$$S(\lambda E, \lambda V, \lambda N) = \lambda S(E, V, N), \tag{1}$$

where λ is a scalar. From the differential of E,

$$dE = T\, dS - P\, dV + \mu\, dN, \tag{2}$$

one gets the differential of S,

$$dS = dE/T + (P/T)\, dV - (\mu/T)\, dN \tag{3}$$

(T is temperature, P the pressure and μ the chemical potential). One sees that the temperature is given by

$$(\partial S/\partial E)_{V,N} = 1/T. \tag{4}$$

The function of state $P(V, T, N)$ is deduced from $T(E, V, N)$ and $P(E, V, N) = T(\partial S/\partial V)_{E,N}$ and eliminating E from them.

2. The Helmholtz free energy as a potential

The second case corresponds to the system in thermal contact with a reservoir which defines the temperature of the system with volume V and number of particles N. It is supposed that: a) The ensemble reservoir with the system constitutes a closed system, b) The reservoir is much larger than the system. The thermodynamic potential is the Helmholtz free energy, which is also a minimum when T, V and N are kept constant. Its definition is

$$F(T, V, N) = E - TS, \tag{5}$$

and the differential of F is

$$dF = dE - d(TS) = T\,dS - P\,dV + \mu\,dN - T\,dS - S\,dT$$

or

$$dF = -S\,dT - P\,dV + \mu\,dN. \tag{6}$$

One gets the entropy

$$S(T, V, N) = -(\partial F/\partial T)_{V,N} \tag{7}$$

and the energy

$$E(T, V, N) = F - T(\partial F/\partial T)_{V,N}. \tag{8}$$

The function of state here is merely $P(T, V, N) = -(\partial F/\partial V)_{T,N}$.

In the case of magnetic material one has to take into account the magnetization of the material under application of a magnetic field. The magnetization M is the number of effective magnetic dipoles and the contribution to the energy is HdM. The differential of E is now

$$dE = T\,dS - P\,dV + \mu\,dV + H\,dM, \tag{9}$$

since E is a function of the extensive quantities as M is. The free energy is $F_M = E - TS - MH$, such that its differential is

$$dF_M = -S\,dT - P\,dV + \mu\,dN - M\,dH. \tag{10}$$

In other words F_M is a function of T, V, N and H. The magnetization is obtained from F_M through its derivative relative to H, $M = -(\partial F_M/\partial H)_{T,V,N}$.

3. The grand potential

In the third case, the system is in thermal contact with a reservoir made of the same kind of particles. But now it is enclosed in a cell of volume V with walls permeable to the particles. The variables which characterize the system are the volume V, the temperature T (imposed by the reservoir) and the chemical potential μ which is the same for the system and the reservoir. If one particle crosses the walls from the reservoir to the system when nothing else is changed, the energy of the reservoir deceases by an amount equal to $-\mu_R$ (chemical potential of the reservoir) when the energy of the system increases by μ_S (chemical potential of the system). Since the total energy has not changed (the reservoir with the system constitutes a closed system), it results in the equality of the chemical potentials.

In the present case the thermodynamic potential is the grand potential

$$\Psi(T, V, \mu) = E - TS - \mu N. \tag{11}$$

From the fundamental equality of thermodynamics, $E = TS - PV + \mu N$ one gets $\Psi = -PV$. The differential of Ψ is $d\Psi = dE - d(TS) - d(\mu N)$, or

$$d\Psi = -S\, dT - P\, dV - N\, d\mu. \tag{12}$$

Thus, one has

$$S = -(\partial \Psi / \partial T)_{V,\mu}, \tag{13}$$

$$P = -(\partial \Psi / \partial V)_{T,\mu}, \tag{14}$$

$$N = -(\partial \Psi / \partial \mu)_{T,V}. \tag{15}$$

The energy is given by

$$E = \Psi + TS + \mu N = \Psi - T(\partial \Psi / \partial T)_{V,\mu} - \mu(\partial \Psi / \partial \mu)_{T,V}. \tag{16}$$

In the first case, the temperature and the pressure are not controlled by some external influence and one considers the system as a closed system. In the second case, the macroscopic energy is the mean value of the energy of the system. It can fluctuate since only the temperature is imposed by the external reservoir. But it is a fundamental hypothesis that the macroscopic energy that can be measured is the mean value and that the fluctuations around the mean value are negligible when the number of particles is very large. It is not always the case but the study of such situations is beyond the framework of this book. And in the third case, the macroscopic energy is the mean energy, and the macroscopic density corresponds to the mean value of N. Since the walls are permeable to the particles, their number in the volume V of the system can vary. If we suppose that the fluctuations are negligible, the three cases give same results and one can use the standard thermodynamics formulas.

The fundamental goal of statistical mechanics is to determine the thermodynamic potentials S, F and Ψ from the knowledge of the microscopic properties of the particles. It is the subject of the two following chapters.

PART I
Fundamentals

Chapter 1

The Closed System or the Microcanonical Ensemble

1.1 The Microcanonical Ensemble

In a closed system, N particles are enclosed in a volume V with walls impervious to heat such that the system cannot receive energy from the external world or send energy outside. The energy of the system is E. We suppose also that the system is in equilibrium. In other words, it is supposed that it is possible to prepare the system with well defined values of the energy E, the volume V, the number of particles N and other quantities which we designate by α. We call such a situation a **macrostate** of the system.

At a particular time, a particular particle is characterized by a set of variables which define completely its state. For example the state of a free and isolated particle is given by the knowledge of its momentum vector $\boldsymbol{p}$. However, the energy is not a characteristic of the state, since there are several states with the same energy and the same absolute value of the momentum $\boldsymbol{p}$, but with different directions of the vector $\boldsymbol{p}$. Also, in a group of several particles, each particle does not stay always in the same state and there is a perpetual change from state to state.

At a particular time, the system is in a state called a **microstate** when the states of all the particles are well defined. We recall the fundamental hypothesis that the particles change their state with time, so the system also passes from a microstate to another, but keeping

9

the same macrostate defined by the chosen values of the quantities E, V and N. If we consider an ensemble (called the microcanonical ensemble) of a very large number of identical systems (prepared as above with the same set of variables E, V, N and α) each system is in a particular microstate when all these microstates correspond to the same macrostate.

We shall give a simple example of a system of three particles which can have one of four possible energies: 0, ε, 2ε or 3ε. We suppose that the total energy of the system is 3ε and this energy defines the macrostate of the system. One can have the following possibilities, each defining a particular microstate: a) the three particles are in the state with energy ε; b) one particle is a the state with energy 3ε and the two others are in the state with energy 0; c) one particle is in the state with energy 2ε, one particle in the state with energy ε and the third in the state with energy 0.

Now we can state several postulates concerning the ensemble.

Postulate 1. The probability p_i, to find a particular system in the ensemble in a given microstate (labeled i), is the same for all the microstates (in number Ω). The probability is a number smaller than 1 such that the sum of all the probabilities is equal to 1. One has $\Sigma p_i = 1$ when the sum has Ω terms. Since p_i is the same for all the microstates, $\Sigma p_i = \Omega p_i = 1$, and one gets $p_i = p = 1/\Omega$.

In other words, there is no preferential microstate for the chosen macrostate. We can by now, emphasize that the number of microstates Ω is a function of the parameters E, V, N and α.

Postulate 2 (Boltzmann). The entropy S of the system is

$$S = k_B \operatorname{Ln} \Omega. \tag{1.1}$$

This is the most fundamental formula of statistical mechanics. Boltzmann first proposed it and it is why the constant k_B is called the Boltzmann constant. Its value is $1.38 \times 10^{-23} \, \mathrm{JK^{-1}}$ or $8.62 \times 10^{-5} \, \mathrm{eVK^{-1}}$. (We recall that $1 \, \mathrm{eV} = 1.602 \times 10^{-19} \, \mathrm{J}$.) We shall see later how this values obtained.

The problem now is to show that this postulate gives the same entropy than that defined in thermodynamics. Later we shall see that we need a new postulate that we formulate below.

An important consequence of this Postulate 2 is that, for $T = 0$, the entropy is null (with some particular exceptions). $T = 0$ is the lowest temperature and all the particles are in their lowest energy state. So the number of microstate is 1 since all the particles are in the same state.

Postulate 3. In a closed system, the equilibrium state corresponds to the largest value of the entropy.

In other words, since E, V and N are fixed, the equilibrium state will be given by the value of α which gives the largest value of S.

Now we shall take two simple examples.

Example 1. One considers a row of six spins aligned. If n_1 spins are up and n_2 spins are down the number of microstates associated with a particular choice of n_1 and n_2 is

$$\Omega = \frac{6!}{n_1! \, n_2!}. \tag{1.2}$$

It is the number of possibilities to put 6 particles in two boxes. If there is no external magnetic field nor interaction between spins, the energy is null. The equilibrium state is the one with the largest number of microstates Ω because in this case the entropy is maximum. In the present case, one can chose as α the number n_1 (or n_2). Varying n_1 one sees that Ω is maximum for $n_1 = 3$ ($\Omega = 20$) as shown in the Table 1.

In the equilibrium state, half of the spins are up and half down. The entropy is equal to $k_B \ln 20$.

Example 2. Three spins are located at the three corners of an equilateral triangle. The energy of the systems is only due to the interactions of the spins. m_1, m_2 and m_3 are the values of the three spins which can be equal to 1 (spin up) or -1 (spin down). The

Table 1. Number of microstates when the number of spin ups is changed.

n_1	0	1	2	3	4	5	6
Ω	1	6	10	20	10	6	1

energy of the system is $E = -(m_1m_2 + m_2m_3 + m_3m_1)$. There are $2 \times 2 \times 2$ microstates depending of the directions of the spin. There are 2 microstates with energy $E_1 = -3$ (the three spins up or the three spins down) and 6 microstates (when two spins are in the same direction and the third is in the opposite) with energy $E_2 = -1$. In other words, the system can have only two macrostates. What are the possible values of the entropy for such system? In the first case, the entropy is $S_1 = k_B \operatorname{Ln} 2$ and in the second case $S_2 = k_B \operatorname{Ln} 6$. Note that the macrostate with larger energy has also the larger entropy.

1.2 Properties of the Entropy

Part A. We consider an isolated system of N particles with energy E in the volume V which is divided in two distinct cells (left cell with volume V_1 and right cell with volume V_2, $V_1 + V_2 = V$). The two cells are separated by a rigid wall covered by a material which does not give the possibility for heat to go from one cell to the other nor for the particles to pass from one side to the other (if it is a gas). In the left side, there are N_1 particles with energy E_1, and N_2 particles with energy E_2 in the right side ($N_1 + N_2 = N$ and $E_1 + E_2 = E$). The two cells are two closed subsystems (Fig. 1.1).

We shall call $\Omega(E, V, N)$ the number of microstates of the whole system, $\Omega_1(E_1, V_1, N_1)$ the number of microstates of the left subsystem and $\Omega_2(E_2, V_2, N_2)$ the number of the microstates of the right subsystem. One has

$$\Omega = \Omega_1 \Omega_2, \tag{1.3}$$

since for each microstate of the left subsystem it is possible to associate all the microstates of the right. We conclude from (1.1) that

$$S = k_B \operatorname{Ln} \Omega = k_B \operatorname{Ln} (\Omega_1 \Omega_2) = k_B \operatorname{Ln} \Omega_1 + k_B \operatorname{Ln} \Omega_2, \tag{1.4}$$

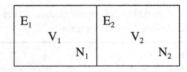

Figure 1.1.   A system and its two subsystems.

or

$$S = S_1 + S_2. \tag{1.5}$$

The entropy is an additive quantity as it must be.

Part B. One considers again the isolated system divided into two subsystems as above. At a given time, one removes the material which covers the separation wall, making it permeable to conduct heat. The two subsystems, being different, were not at the same temperature, but now that the wall is permeable to heat, there is heat transfer from the side with higher temperature to the side with lower temperature. For $t = \infty$ the two sides reach the same temperature. At $t = 0$, when one removes the isolating material, the system is not in equilibrium, and reaches it at the end of the process. At the equilibrium, the entropy reaches its largest value.

To understand this point one writes (with V and N constant)

$$S = S_1(E_1) + S_2(E_2) = S_1(E_1) + S_2(E - E_1) \tag{1.6}$$

since the total energy E is constant. During the heat transfer process, one can take E_1 as the quantity α which maximizes S. This condition is

$$(\partial S/\partial E_1)_{V,N} = (\partial S_1(E_1)/\partial E_1)_{V,N} + (\partial S_2(E - E_1)/\partial E_1)_{V,N} = 0 \tag{1.7}$$

or

$$(\partial S/\partial E_1)_{V,N} = (\partial S_1/\partial E_1)_{V,N} + (\partial S_2/\partial E_2)_{V,N}(dE_2/dE_1) = 0. \tag{1.8}$$

But $dE_2/dE_1 = -1$, since $E_2 = E - E_1$.

The condition $(\partial S/\partial E_1)_{V,N} = 0$ which gives the condition of equilibrium can be written

$$(\partial S_1/\partial E_1)_{V,N} = (\partial S_2/\partial E_2)_{V,N}. \tag{1.9}$$

Since the equilibrium is given by the equality of the temperatures $T_1 = T_2$, one concludes that the equality (1.9) is equivalent to the equality of the temperatures. It means that the derivatives $(\partial S_i/\partial E_i)_{V,N}$ are functions of T only. Now we add the last postulate:

Postulate 4. The derivative of the entropy relative to the energy is equal to the inverse of the temperature

$$(\partial S/\partial E)_{V,N} = 1/T. \tag{1.10}$$

Equation (1.10) gives the link with the thermodynamic definition of the entropy. It is not difficult to see that the derivative $(\partial S/\partial E)_{V,N}$ is positive: if one increases the energy of the system, keeping V and N constant, the possible energies for each particle is also increased giving more possible states to it. This means that the total number of microstates has also been increased.

It remains to show that in the process to equalize the temperatures there is an increase of the entropy. It is a well known result of thermodynamics that an irreversible process takes place with increasing of the entropy. For this result we compare the entropy S_i at the beginning of the process with the entropy S_f at its end.

At the beginning the number of microstates is given by (as shown above)

$$\Omega_i = \Omega_1(E_1)\Omega_2(E - E_1). \tag{1.11}$$

However in the final state, the energy E_1 of subsystem one is not fixed, but can take several values (due to fluctuations in the temperatures) keeping the sum $E_1 + E_2$ constant. For the calculation of the number of microstates we have to sum over all the possible values E_i of subsystem. We get

$$\Omega_f = \sum \Omega(E_1)\Omega(E - E_1). \tag{1.12}$$

In the expression (1.12), there is a term in the sum with E_i equal to the initial energy E_1 and other similar terms. This means that the sum (1.12) is larger than (1.11). In other words, the number of microstates in the final state is larger than in the initial state. We conclude that the entropy increased in the process of equalization of temperatures.

Part C. One considers the same system as above, divided into two subsystems. At the beginning the wall is clamped such the pressures on both sides are not equal. However at a given time, the wall is freed and moves until the pressures on both sides are equal. At the same time, the wall is also made permeable to transfer heat so the temperature on both sides will be equal. Taking V_1 and E_1 as quantities, for which one looks for equilibrium by maximizing the entropy, we write $\partial S/\partial E_1 = 0$ and $\partial S/\partial V_1 = 0$. One gets the following equalities (following the same method as above)

$$\left(\frac{\partial S_1}{\partial E_1}\right)_{V,N} = \left(\frac{\partial S_2}{\partial E_2}\right)_{V,N} \tag{1.13}$$

and

$$\left(\frac{\partial S_1}{\partial V_1}\right)_{E,N} = \left(\frac{\partial S_2}{\partial V_2}\right)_{E,N}, \tag{1.14}$$

which are equivalent to stating the equality of the pressures and the temperatures. We have already seen that (1.13) is equivalent to $T_1 = T_2$. To translate the equivalence of (1.14) we need to take into account the dimensionality of the ratio (entropy/volume). From (1.10), one sees that the dimensionality of S is (energy/temperature) or (volume $\times$ pressure/temperature). Thus the dimension of the ratio (entropy/volume) is (pressure/temperature). We can replace the equality (1.14) by the following equality:

$$\frac{P_1}{T_1} = \frac{P_2}{T_2}, \tag{1.15}$$

with the conclusion that

$$\left(\frac{\partial S}{\partial V}\right)_{E,N} = \frac{P}{T}. \tag{1.16}$$

Part D. Once again, one uses the system divided into two subsystems. Up to now, the physical state of the system (solid, liquid or gas) was not important since the number of particles N_1 and N_2 were constant. Now we suppose that it is a fluid with the wall not permeable to particles. At a given time, one makes the wall a) permeable to heat, b) mobile, and c) permeable to particles. The equilibrium is reached when a) the temperatures are equal, b) the pressures are equal, and c) when the chemical potentials of both sides are equal, that is

$$\mu_1 = \mu_2. \tag{1.17}$$

To see the result we repeat again the argument presented above. Consider the system at equilibrium when some particles cross the wall (although the mean number of particles in both sides is constant, there is always the possibility for particles to cross the wall and make the numbers at various instances different from the mean values). In subsystem 1, the number of particles increases by dN_1, and in the other subsystem by dN_2 with $dN_1 + dN_2 = 0$. The changes in the energies of the two sides are respectively $dE_1 = \mu_1\, dN_1$ and $dE_2 = \mu_2\, dN_2$. Taking into account that the total energy is constant $(dE_1 + dE_2 = 0)$, and that $dN_1 = -dN_2$, one gets the equality (1.17).

We write again the condition of equilibrium for maximizing the entropy selecting as the α variables E_1, V_1 and N_1. Proceeding as above, one gets the equalities (1.13), (1.14) and a new equality

$$\left(\frac{\partial S_1}{\partial N_1}\right)_{E,V} = \left(\frac{\partial S_2}{\partial N_2}\right)_{E,V}, \tag{1.18}$$

which is equivalent to (1.17). Taking into account that the dimensionality of S is (energy/temperature), that μ is an energy and the fact that N is without physical dimension (a scalar), one obtains

$$\left(\frac{\partial S}{\partial N}\right)_{E,V} = -\frac{\mu}{T}. \tag{1.19}$$

The presence of the negative sign is nontrivial and can be justified using the theorem which gives the relation between the three partial derivatives of three quantities related by a relationship. Let be these three quantities x, y and z, and the relation can be written in three equivalent forms: $x(y,z)$, $y(x,z)$ or $z(x,y)$. It is possible to show that (this appears often in textbooks of thermodynamics)

$$\left(\frac{\partial x}{\partial y}\right)_z \left(\frac{\partial y}{\partial z}\right)_x \left(\frac{\partial z}{\partial x}\right)_y = -1. \tag{1.20}$$

We shall use this relation for the entropy seen as a function of E and N, keeping V as a constant and we get

$$\left(\frac{\partial S}{\partial E}\right)_{N,V} \left(\frac{\partial E}{\partial N}\right)_{S,V} \left(\frac{\partial N}{\partial S}\right)_{E,V} = -1, \tag{1.21}$$

or, introducing $(\partial S/\partial E)_{N,V} = 1/T$ and the definition of $\mu = (\partial E/\partial N)_{S,V}$ giving

$$\frac{1}{T}\mu \left(\frac{\partial N}{\partial S}\right)_{E,V} = -1, \tag{1.22}$$

from which one deduces (1.19). Without the negative sign, it should not be possible to establish the important equation (1.20)

From the expressions (1.10), (1.16) and (1.19), we get the differential of the entropy

$$dS = \frac{dE}{T} + \frac{P}{T}dV - \frac{\mu}{T}dN. \tag{1.23}$$

1.3 An Example

We now solve a system of N particles in a solid with two possible states when one supposes that N is very large. Each state is characterized by its energy: state 1 with energy 0 and state 2 with energy e. In the present case, the volume is constant and does not play any role, so we drop it. The number of particles being constant, the different interesting quantities are functions of only T. If one considers the system as closed, it is possible to apply the method of the microcanonical ensemble. In a given macrostate, the number of particles in each state is specified, and this corresponds to a particular value of the energy and also to the temperature. Taking a different macrostate (i.e. a different distribution of the particles between the two possible states) one gets a different value of the energy and temperature. The question is to find the energy, entropy and number of particles in each particular state as functions of the temperature. First one calculates the entropy.

If n particles are in the state 2, the energy of the system is $E = ne$. The number of microstates is given by

$$\Omega = \frac{N!}{n!(N-n)!}. \tag{1.24}$$

It is the number of ways to chose n particles among N particles and is equal to the number of ways to put N particles in two boxes, n particles in the first and $N - n$ particles in the second. The entropy is

$$S = k_B \, \mathrm{Ln} \, \Omega = k_B [\mathrm{Ln} \, N! - \mathrm{Ln} \, n! - \mathrm{Ln}(N-n)!]. \tag{1.25}$$

To go further, we adopt Stirling's approximation $\mathrm{Ln} \, N! \approx N \, \mathrm{Ln} \, N - N$, which is accurate, even for a relatively small number, say 10^4. In the present case of a macroscopic body, n and N may be vary large (in the order of 10^{19}) and, as we shall see below, even when the approximation used is not valid the results are qualitatively correct.

From (1.25), one gets

$$S = k_B[N \operatorname{Ln} N - n \operatorname{Ln} n - (N - n) \operatorname{Ln}(N - n)]. \tag{1.26}$$

In order to find the functions $E(T)$, $S(T)$ and $n(T)$, one writes the relation $(\partial S/\partial E) = 1/T$ using $dS = (\partial S/\partial n)_E \, dn$ and $dE = e \, dn$. This gives (by writing $dn = dE/e$)

$$\left(\frac{\partial S}{\partial E}\right)_n = \left(\frac{\partial S}{\partial n}\right)_E \left(\frac{\partial n}{\partial E}\right) \tag{1.27}$$

or

$$\left(\frac{\partial S}{\partial E}\right)_n = \left(\frac{1}{e}\right)\left(\frac{\partial S}{\partial n}\right)_E = 1/T. \tag{1.28}$$

From (1.26) one calculates the derivative $(\partial S/\partial n)_E$ and the result following (1.28), that is $(\partial S/\partial n)_E = e/T$,

$$\left(\frac{\partial S}{\partial n}\right)_E = k_B \operatorname{Ln}\left(\frac{N - n}{n}\right) = e/T. \tag{1.29}$$

From (1.29), one can extract the relation between n and T

$$n = \frac{N}{1 + \exp\left(\frac{e}{k_B T}\right)}. \tag{1.30}$$

Now it is easy to insert into the expressions of the energy and entropy $n(T)$. One has that for E,

$$E = ne = \frac{Ne}{1 + \exp(\frac{e}{k_B T})}, \tag{1.31}$$

and that for S,

$$S = k_B N\left[\left(1 + \exp\left(\frac{e}{k_B T}\right)\right)^{-1} \operatorname{Ln}\left(1 + \exp\left(\frac{e}{k_B T}\right)\right)\right.$$
$$\left. + \left(1 + \exp\left(-\frac{e}{k_B T}\right)\right)^{-1} \operatorname{Ln}\left(1 + \exp\left(-\frac{e}{k_B T}\right)\right)\right],$$

$$\tag{1.32a}$$

which can also be written in the following form

$$S = k_B N \left(\frac{e}{k_B T} \right) \left[1 + \exp\left(\frac{e}{k_B T} \right) \right]^{-1}$$
$$+ k_B N \, \text{Ln} \left[1 + \exp\left(-\frac{e}{k_B T} \right) \right]. \qquad (1.32b)$$

The two expressions (1.32a) and (1.32b) look so different such that one might doubt if it is really the same function. It is left as an exercise for the reader to find the way to pass from one expression to another. We suggest to use the following identities: $(1 + 1/x)^{-1} + (1 + x)^{-1} = 1$ and $\text{Ln}\frac{1+e^x}{1+e^{-x}} = x$. The complete derivation is given at the end of the chapter.

Now it is interesting to derive for the limits of n and S at low and high temperatures. The low temperatures are defined by the condition $k_B T \ll e$ or $\frac{e}{k_B T} \gg 1$. This gives for n (since $\frac{e}{k_B T}$ goes to infinity and $\exp(\frac{e}{k_B T}) \to \infty$),

$$n = N \exp\left(-\frac{e}{k_B T} \right), \qquad (1.33)$$

and one sees that n goes to zero as T goes to zero.[1] This result was expected since for $T = 0$ all the particles are in the state one with the lowest energy, and n goes to 0. So the total energy is also zero.

For the entropy, one has to include only the first term of (1.32b) (the second term goes to 0 because $\exp(-\frac{e}{k_B T})$ goes to 0) and this first term becomes

$$S = k_B N \left(\frac{e}{k_B T} \right) \exp\left(-\frac{e}{k_B T} \right), \qquad (1.34)$$

which goes to zero for $T = 0$ (recalling that the product $x \exp(-x)$ goes to zero if x goes to ∞) as expected since there is only one microstate.

Now at high temperatures, when $k_B T \gg e$, one obtains (since $\frac{e}{k_b T} \ll 1$, one can use the approximations valid for $x \ll 1$, that

---

[1] In this case, the Stirling approximation is not always applicable. Nevertheless, the result is qualitatively correct.

$\exp(x) \approx + x$ and $1/(1 + x) \approx 1 - x$)

$$n = \frac{N}{(2 + \frac{e}{k_B T})} = \frac{N}{2}\left[1 - \frac{e}{2k_B T}\right] \tag{1.35}$$

and

$$S = Nk_B\left[\text{Ln } 2 + \left(\frac{e}{2k_B T}\right)^2\right]. \tag{1.36}$$

As T goes to infinity, n tends to $N/2$, there is equal distribution of particles between the two states, and S tends to Nk_B Ln 2.

In Fig. 1.2, we give the variations of $\frac{n}{N}$ and $\frac{S}{k_B N}$ with T when one chooses $e/k_B = 10$.

Important remark. In the determination of the number of micro-states, one admits that it is possible to follow separately each particle. Consequently, when there are n particles in level 2, if one permutes one particle form level 1 with one particle from level 2, one has the same energy but a different microstate. We say that the particles are distinguishable. We did not introduce the quantum mechanical

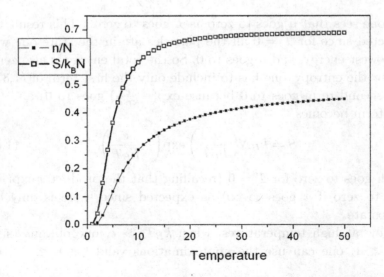

Figure 1.2.   Variations of the entropy and the relative number of particles in the state with the largest energy as functions of the temperature. The ratio n/N tends to 0.5 when the ratio $(S/k_B N)$ tends to Ln 2 $\sim$ 0.7.

behavior of the particles, thus in this example, the particles are classical and can be distinguished.

Equivalence of (1.32a) and (1.32b).

Writing $x = \frac{e}{k_B T}$, (1.32) can be written as

$$\frac{S}{k_B N} = \frac{\mathrm{Ln}(1 + e^x)}{1 + e^x} + \frac{\mathrm{Ln}(1 + e^{-x})}{1 + e^{-x}}. \tag{1.37}$$

Using the identity $\mathrm{Ln}\frac{1+e^x}{1+e^{-x}} = x$ one can write

$$\mathrm{Ln}(1 + e^x) - \mathrm{Ln}(1 + e^{-x}) = x,$$

or

$$\mathrm{Ln}(1 + e^x) = x + \mathrm{Ln}(1 + e^{-x}). \tag{1.38}$$

Introducing this expression of $\mathrm{Ln}(1 + e^x)$ in (1.37), one gets

$$\frac{S}{k_B N} = \frac{x + \mathrm{Ln}(1 + e^{-x})}{1 + e^x} + \frac{\mathrm{Ln}(1 + e^{-x})}{1 + e^{-x}} \tag{1.39}$$

One sees that one can develop (1.39) and regroup the terms as

$$\frac{S}{k_B N} = \frac{x}{1 + e^x} + \mathrm{Ln}(1 + e^{-x})\left[\frac{1}{1 + e^x} + \frac{1}{1 + e^{-x}}\right]. \tag{1.40}$$

Now one uses the second identity $1/(1 + e^x) + 1/(1 + e^{-x}) = 1$ and gets

$$\frac{S}{k_B N} = \frac{x}{1 + e^x} + \mathrm{Ln}(1 + e^{-x}), \tag{1.41}$$

which is (1.32b) when x is replaced by $\frac{e}{k_B T}$.

Chapter 2

The System in Thermal Contact with a Reservoir: The Canonical and Grand Canonical Ensembles

In this chapter, we shall study a system in contact with a very large system which stays at constant temperature T. We shall call it the reservoir. In the first case, the contact is purely thermal; it means that the temperature of the system is fixed by contact with the reservoir. In the second case, the contact is thermal but there is also the possibility of exchange of particles between the system and the reservoir. Since it is supposed that the system and the reservoir are in equilibrium, there are equalities of the temperatures and of the chemical potentials, as explained above. Our goal is always to find the relation between the microscopic properties of the particles of the system and its macroscopic properties.

The fundamental hypotheses are:

(a) The system and the reservoir are seen together as a grand system which is isolated from the rest of the world. The results of the preceding chapter can be applied to the grand system. In particular, its energy E_0 is constant and the entropy is given by (1.1).

(b) The system is much smaller than the reservoir itself. In particular the energy E_S of the system is much smaller than the energy E_R of the reservoir and that of the grand system. Thus one will be able to apply approximations concerning small quantities versus large ones.

2.1 The Canonical Ensemble

2.1.1 *The partition function*

As revealed above, the system with volume V and number of particles N is in thermal contact with the reservoir, which imposes on the system the temperature. However, the energy of the system is not fixed by the contact with the reservoir and can fluctuate over all the possible energies of the system. Our first goal will be to determine, in the ensemble of a huge number of identical systems which constitutes the canonical ensemble, the probability to find a system in a microstate labeled s with energy E_S. Once this probability is known, we shall be able to calculate the mean energy, which is the macroscopic energy of the system and other quantities.

If one picks at random a system in the ensemble, the probability p_s to find this system in a microstate s with energy E_S is the ratio of two quantities: the number of microstates of the system with energy E_S against the total number of microstates of the grand system. A probability is defined as the number of "favorable" cases (here a favorable case is the system is a microstate with energy E_S) divided by the number of all the possible cases (here the number of microstates of the grand ensemble, irrespective they are favorable or not).

If the energy of the system is E_S and that of the grand system is E_0, the energy of the reservoir is $E_0 - E_S$ and the number of microstates in which the system has energy E_S is equal to the number of microstates in which the reservoir has the energy $E_0 - E_S$. One writes

$$p_s = \frac{\Omega_R(E_0 - E_S)}{\Omega_{GS}(E_0)}, \tag{2.1}$$

where Ω_R indicates the number of microstates of the reservoir and Ω_{GS} those of the grand system. From the previous chapter, we know that the entropies S_{GS} of the grand system and that S_R of the reservoir are respectively

$$S_{GS} = k_B \operatorname{Ln} \Omega_{GS}(E_0),$$

$$S_R = k_B \operatorname{Ln} \Omega_R(E_0 - E_S),$$

and we have

$$\Omega_{GS}(E_0) = \exp[S_{GS}(E_0)/k_B], \tag{2.2}$$

$$\Omega_R(E_0 - E_s) = \exp[S_R(E_0 - E_s)/k_B]. \tag{2.3}$$

For the grand system there is no problem in applying the preceding formula since it is a closed system. For the reservoir, it is only an approximation since it is in contact with the system and is not a closed system. It is a good approximation since it is much larger than the system and is practically unperturbed by the system.

From (2.1)–(2.3) one gets

$$p_s = \frac{\exp[S_R(E_0 - E_S)/k_B]}{\exp[S_{GS}(E_0)/k_B]}. \tag{2.4}$$

One can expand $S_R(E_0 - E_S)$ as a Taylor expansion[1] and terminate after the second term since $E_0 \gg E_S$, and

$$S_R(E_0 - E_S) = S_R(E_0) - E_S(\partial S_R/\partial E)_0, \tag{2.5}$$

but $(\partial S_R/\partial E) = 1/T$,

$$S_R(E_0 - E_s) = S_R(E_0) - E_S/T. \tag{2.6}$$

The probability p_s can be written as

$$p_s = \frac{\exp[S_R(E_0)/k_B - E_S/(k_B T)]}{\exp[S_{GS}(E_0)/k_B]}$$

or

$$p_s = \frac{\exp[S_R(E_0)/k_B] \exp[-E_S/(k_B T)]}{\exp[S_{GS}(E_0)/k_B]}. \tag{2.7}$$

Since the sum of all the probabilities is equal to 1, equivalently

$$\sum_S p_s = \sum_S \frac{\exp[S_R(E_0)/k_B] \exp[-E_S(k_B T)]}{\exp[S_{GS}(E_0)/k_B]} = 1 \tag{2.8}$$

---

[1]The Taylor expansion of a function $f(x)$ is given by $f(x+h) = f(x) + h\,df/dx(x) + (h^2/2!)\,d^2f/dx^2(x) + \cdots + (h^n/n!)\,d^nf/dx^n(x) + \cdots$.

or

$$\sum_S p_s = \frac{\exp[S_R(E_0)/k_B]}{\exp[S_{GS}(E_0)/k_B]} \sum_S \exp\left[-\frac{E_S}{k_B T}\right] = 1,$$

thus,

$$\exp[S_{GS}(E_0)/k_B] = \exp[S_R(E_0)/k_B] \sum_S \exp\left[-\frac{E_S}{k_B T}\right]. \tag{2.9}$$

Substituting (2.9) into (2.7), we get the final result for p_s:

$$p_s = \exp\left[-\frac{E_S}{k_B T}\right] \Big/ \sum_S \exp\left[-\frac{E_S}{k_B T}\right], \tag{2.10}$$

when the sums in (2.8)–(2.10) are over all the possible microstates and not only on the possible values of the energies. The probability p_s is the probability to find the system in a microstate or shortly in a state with energy E_S. It is an important result and we shall use it several times.

The sum

$$Z = \sum_S \exp\left[-\frac{E_S}{k_B T}\right] \tag{2.11}$$

is the called the partition function. In the context of the canonical ensemble, this is the most important result. We shall see that it will give the link between the microscopic and the macroscopic points of view. Z is a function of the parameters T, V, and N, that define the system. The temperature T appears explicitly, but the possible energies of the system depend in general on V and N.

It is common to take the Greek letter β as the inverse of the temperature $\beta = 1/(k_B T)$ and to write Z as

$$Z = \sum_s \exp(-\beta E_S). \tag{2.12a}$$

It is possible that several different microstates have the same energy, thus there is a function $g(E_S)$, giving the number of microstates for a given energy E_S. Such states with the same energy are called degenerate states. The partition function can be written as a sum over the energies:

$$Z = \sum_E g(E_S) \exp(-\beta E_S). \tag{2.12b}$$

We stress the difference between, (2.11) or (2.12a), and (2.12b). In (2.11) the sum is over all the microstates and in (2.12b) it is over the energies.

2.1.2 *Energy, entropy and thermodynamic potential*

Part A. With the knowledge of p_s, it is possible to calculate the mean energy E as

$$E = \sum_S p_s E_S, \qquad (2.13a)$$

or

$$E = \frac{\sum_S E_S \exp(-\beta E_S)}{\sum_S \exp(-\beta E_S)}. \qquad (2.13b)$$

The formula (2.13a) is the standard formula for the mean value of a quantity which has different probability to appear in different states of respective energies.

The numerator of (2.13b) is minus the derivative of the dominator relatively to β, and the energy can written, using Z, as

$$E = -\left[\frac{\partial Z}{\partial \beta}\right]\Big/ Z = -\frac{\partial \operatorname{Ln} Z}{\partial \beta}. \qquad (2.14)$$

Part B. The energy E is given as a function of T, V and N. The derivative $(\partial E/\partial T)_{V,N}$ is the specific heat at constant volume, C_V. The entropy is given by[2] $S = \int C_V(dT/T) = \int (\partial E/\partial T)_V \, (dT/T)$. One uses the following relations linking T and β to get:

$$T = \frac{1}{k_B \beta},$$

$$dT = -\frac{d\beta}{k_B \beta^2},$$

$$\frac{\partial E}{\partial T} = \frac{\partial E}{\partial \beta}\frac{d\beta}{dT} = -k_B \beta^2 \frac{\partial E}{\partial \beta},$$

giving $S = \int(\partial E/\partial T)_V (dT/T)k_B = k_B \int \beta \, (\partial E/\partial \beta)_V \, d\beta$.

---

[2]The fundamental relation between the entropy dS and the heat dQ in an infinitesimal process is $dS = dQ/T$. If the process takes place at constant volume $dQ = C_V \, dT$ and $dS = (C_V/T) \, dT$.

Performing integration by parts, one writes

$$u = \beta, \qquad dv = \left(\frac{\partial E}{\partial \beta}\right)_V d\beta, \qquad \int u\,dv = uv - \int v\,du,$$

$$du = d\beta, \qquad v = E.$$

One gets, using (2.14), that

$$S = k_B\left[\beta E - \int E\,d\beta\right] = k_B\left[\beta E + \int \frac{\partial \mathrm{Ln}\,Z}{\partial \beta}d\beta\right], \qquad (2.15)$$

or

$$S = E/T + k_B\,\mathrm{Ln}\,Z. \qquad (2.16)$$

From (2.16) we find the link that we are looking for:

$$E - TS = F = -k_B T\,\mathrm{Ln}\,Z. \qquad (2.17)$$

We recall that F is the Helmholtz free energy and a function of T, V and N. We mentioned above that from the knowledge of F we can get all the possible data about the system. The relation (2.17) is one of the most important in this course of study.

We close this section by another formulation of the entropy:

$$S = -k_B \sum_S p_s\,\mathrm{Ln}\,p_s. \qquad (2.18a)$$

To verify, one introduces in (2.18a) the expression (2.10) for p_s and one obtains

$$S = -k_B \sum_S [-\beta E_S - \mathrm{Ln}\,Z]\frac{\exp(-\beta E_S)}{Z}, \qquad (2.18b)$$

or

$$S = k_B \sum_S [\beta E_S \exp(-\beta E_S)]/Z + k_B[\mathrm{Ln}\,Z/Z]\sum_S \exp(-\beta E_S).$$

$$(2.18c)$$

Taking into account (2.13b), one sees that the first term in (2.18b) is equal to E/T. Since $\sum_S \exp(-\beta E_S) = Z$, the second term becomes equal to $k_B\,\mathrm{Ln}\,Z$. Finally (2.18c) is equal to $E/T + k_B\,\mathrm{Ln}\,Z$. One recovers the expression (2.16) for the entropy.

The expression (2.18a) is very general and can be used for the closed system of the microcanonical ensemble. In this case, the probability to find the system in a microstate with the chosen energy is

$1/\Omega$, since all the microstates, in number Ω, have the same probability to be found in the ensemble. Putting $p_s = 1/\Omega$ in (2.17) gives again the Boltzmann formula (1.1), $S = k_B \operatorname{Ln} \Omega$ since the number of terms in the sum is merely Ω.

Part C. We consider N particles without interaction in a volume V at temperature T. We add also that it is possible to distinguish between the particles. This implies that it is always possible (in principle) to follow an individual particle. In such case, the permutation of two particles between their respective states introduces a new microstate for the system. This cannot be true for gas in which there are constant collisions between the particles such that their "individuality" is lost.

In the expression of Z will appear all the energies of the system. These energies are all the possible sums of the individual energies (of the microstates)[3] of the N particles. We note by $\{e_i\}$ one of these possible sums (with N terms), and the partition function is $Z = \sum \exp(-\beta\{e_i\})$ over all these $\{e_i\}$.

The one-particle partition function is $Z_1 = \sum \exp(-\beta e_i\}$, where the sum is on the possible energies of one particle. Now consider the following expression B:

$$B = (Z_1)^N = [\exp(-\beta \sum_i e_i)]^N. \qquad (2.19)$$

One can write

$$B = [\exp(-\beta \sum_i e_i)]^N$$

$$= [\exp(-\beta \sum_i e_i)][\exp(-\beta \sum_i e_i)][\exp(-\beta \sum_i e_i)] \ldots \qquad (2.20a)$$

when the right hand contains N identical terms. More explicitly (2.20a) is

$$[\exp(-\beta e_1) + \exp(-\beta e_2) + \exp(-\beta e_3)\ldots][\exp(-\beta e_1) + \exp(-\beta e_2)$$

$$+ \exp(-\beta e_3)\ldots][\exp(-\beta e_1) + \exp(-\beta e_2) + \exp(-\beta e_3)\ldots].$$

$$(2.20b)$$

---

[3]One can sort the energies of the microstates of one particle as $e_1, e_2, e_3, \ldots$ The energies e_i are not necessarily all different since it is possible that some microstates have the same energy.

If one expands these products, one finds that B will be given by a sum of terms of the form $\exp[-\beta\{e_i\}]$. As above $\{e_i\}$ mentions one of the possible sums of the energies of the particles. This gives

$$B = \sum \exp[-\beta\{e_i\}],$$

and one sees that B is the partition function of the system. Consequently,

$$Z = (Z_1)^N. \tag{2.21}$$

We shall take a simple example of a system with two particles, each with two possible energies e_1 and e_2. The possible energies of the system or the different $\{e_i\}$ are: $e_1+e_1, e_1+e_2, e_2+e_1$ and e_2+e_2. One has

$$Z = \exp[-\beta(e_1 + e_1)] + \exp[-\beta(e_1 + e_2)]$$
$$+ \exp[-\beta(e_2 + e_1)] + \exp[-\beta(e_2 + e_2)]. \tag{2.22}$$

or noticing that $\exp[-\beta(e_1+e_1)] = \exp(-\beta e_1)\exp(-\beta e_1)$ one has, for $a_i = \exp(-\beta e_i)$, that $Z = a_1 a_1 + 2a_1 a_2 + a_2 a_2 = (a_1 + a_2)^2 = (Z_1)^2$.

2.1.3 *A two-level system*

We take the same example that we solved at the end of the preceding chapter. We recall that there are, for each particle, two possible states, one with energy 0 and the other with energy e.

We determine the partition function by the following two ways. The first is to calculate Z for the system of N particles. The second is to use (2.21) and to calculate the one particle partition function.

To be able to compute the partition function, we need to know the number of microstates with a given energy $E = ne$. This number has already been calculated and is equal to $C_{nN} = N!/[(N-n)!n!]$. Thus Z is using (2.12b),

$$Z = \sum_n C_{nN} \exp(-\beta n e) = \sum_n \frac{N!}{(N-n)!n!} x^n, \tag{2.23}$$

with

$$x = \exp(-\beta ne).$$

The sum in (2.23) is the development of the quantity $(1 + x)^N$. Thus the final result is

$$Z = [1 + \exp(-\beta e)]^N. \tag{2.24}$$

This result agrees with what we said about the partition function of independent particles, $Z = Z_1^N$. The one-particle partition function is $Z_1 = 1 + \exp(-\beta e)$, since for a single particle there are only two states.

The free energy F is $-k_B T \operatorname{Ln} Z = -k_B TN \operatorname{Ln}[1 + \exp(-\beta e)]$. Now it is straightforward to get the energy and the entropy using the formulas (2.14) and (2.16) or from the derivatives of F, that they are $E = F - T(\partial F/\partial T)_N$ and $S = -(\partial F/\partial T)_N$. One gets

$$E = \frac{Ne}{1 + \exp(\beta e)},$$

and

$$S = k_B N \left[\frac{\frac{e}{k_B T}}{\left(1 + \exp\left(\frac{e}{k_B T}\right)\right)} + \operatorname{Ln}\left(1 + \exp\left(-\frac{e}{k_B T}\right)\right) \right],$$

as founded above. (Eqs. (1.31) and (1.32b)).

Finally one calculates the chemical potential $\mu = (\partial F/\partial N)_{T,V}$:

$$\mu = \frac{\partial}{\partial N}\{-k_B TN \operatorname{Ln}[1 + \exp(-\beta e)]\} = -k_B T \operatorname{Ln}[1 + \exp(-\beta e)].$$

$$\tag{2.25}$$

2.1.4   *The ideal gas; equipartition of energy in classical mechanics*

In this section we consider particles in classical mechanics. The energy of one particle is the sum of two terms, the kinetic energy E_K and the potential energy V_p, when E_K is a quadratic function of the particle velocities or the angular momentums and V is function of the particle positions. In one dimension (particles on a line of length L) the kinetic energy of one particle is $E_K = p_x^2/2m$ and the partition function is

$$Z = \frac{1}{Q} \iint \exp[-\beta(E_K + V_p)]\, dp_x\, dx, \qquad (2.26)$$

where the lower and upper limits of the integrals for the two variables (p_x, x) are respectively $-\infty$ and ∞ for the momentum, and 0 and L for the position. It is the application of the general expression of the partition function (2.11)

$$Z = \sum_S \exp\left[-\frac{E_S}{k_B T}\right]$$

to the case of a single particle in classical mechanics. The introduction of the quantity Q is needed, since by definition Z is a quantity without dimension, but the double integral in (2.26) has the dimensionality of (momentum)(length). This is the reason why one has to introduce the quantity Q with this dimension. In fact the dimensionality of the product (momentum)(length) is the product (mass)(velocity)(length) or $[M][v][L]$. This product has the dimension of (energy)(time) $= [E][T]$. One can see by the following transformations $[M][v][L] = [M[v][L][T]/[T] = [M][v]^2[T] = [E][T]$. We shall see below which constant was chosen for Q.

The expression (2.26) can be transformed in the product of two integrals

$$Z = (1/Q) \int_{-\infty}^{\infty} \exp\left[-\frac{\beta p_x^2}{2m}\right] dp_x \int_0^L \exp[-\beta V_p(x)]\, dx. \qquad (2.27)$$

Now we consider the case of particles without potential energy and only with kinetic energy. It is the case of the monatomic ideal gas where the atoms have only kinetic energy and there is no interaction between them. This absence of interaction is characteristic of the concept of ideal gas. The one-particle partition function becomes

$$Z_1 = \frac{1}{Q} \int_{-\infty}^{\infty} \exp\left[-\frac{\beta p_x^2}{2m}\right] dp_x \int_0^L dx$$

$$= \frac{L}{Q} \int_{-\infty}^{\infty} \exp\left\{-\left[\frac{p_x^2}{2m\, k_B T}\right]\right\} dp_x.$$

To calculate Z_1 we use the following trick: we divide the integral by $(2m\,k_BT)^{0.5}$ and multiply it by the same factor. This gives

$$Z_1 = (2m\,k_BT)^{0.5}\frac{L}{Q}\int_{-\infty}^{\infty}\exp\left\{-\left[\frac{p_x^2}{2m\,k_BT}\right]\right\}d[p_x/(2m\,k_BT)^{0.5}].$$

(2.28)

The integral is now a definite integral of the variable $u = p_x/(2m\,k_BT)^{0.5}$ with the same limits ($-\infty$ and ∞)

$$Z_1 = (2m\,k_BT)^{0.5}\frac{L}{Q}\int_{-\infty}^{\infty}\exp(-x^2)\,dx.$$

If there is no potential energy for the particle the partition function becomes

$$Z_1 = A(2m\,k_BT)^{0.5}\frac{L}{Q}$$

(2.29)

where $A = \int_{-\infty}^{\infty}\exp(-x^2)\,dx = \sqrt{\pi}$. The energy is calculated through the formulas $E = F - T\frac{\partial F}{\partial T}$ and $F = -k_BT\,\mathrm{Ln}\,Z_1$. Note that we do not need to need the value of A to calculate the energy. One gets

$$E = \frac{1}{2}k_BT.$$

(2.30)

This energy is the mean kinetic energy of the particle and is called the thermal energy of the particle. It is a remarkable result that this mean kinetic energy is proportional to the temperature. For N particles one multiplies the above result by N.

In the case of a particle in three dimensions the kinetic energy is $E_K = (p_x^2 + p_y^2 + p_z^2)/2m$. The partition function is now a multiple integral

$$Z_1 = \left(\frac{1}{Q^3}\right)\int\exp\left\{-\beta\left[\frac{p_x^2+p_y^2+p_z^2}{2m}+V_p(x,y,z)\right]\right\}dp_x\,dp_y\,dp_z\,dx\,dy\,dz.$$

(2.31)

If there is no potential energy it reduces to

$$Z_1 = \left(\frac{1}{Q^3}\right)\int\exp\left\{-\beta\left[\frac{p_x^2+p_y^2+p_z^2)}{2m}\right]\right\}dp_x\,dp_y\,dp_z\,dx\,dy\,dz.$$

(2.32)

This can written as the product of three integrals (when the integral on the variables x, y and z is equal to V)

$$Z_1 = \left(\frac{V}{Q^3}\right) \int_{-\infty}^{\infty} \exp\left[-\frac{\beta p_x^2}{2m}\right] dp_x \int_{-\infty}^{\infty} \exp\left[-\frac{\beta p_y^2}{2m}\right] dp_y$$

$$\times \int_{-\infty}^{\infty} \exp\left[-\frac{\beta p_z^2}{2m}\right] dp_z, \tag{2.33}$$

and from the above results we conclude

$$Z_1 = \frac{V}{Q^3}(2\pi m \, k_B T)^{3/2}, \tag{2.34}$$

and finally

$$E = \frac{3}{2}k_B T. \tag{2.35}$$

The preceding results can be generalized in the following formulation. If in the total energy of a system there is a term which is quadratic in some parameter like the momentum or the position, this term contributes to the energy by the amount $(1/2)k_B T$. The parameter is called a degree of freedom.

A simple application of this theorem of equipartition is the one-dimensional harmonic oscillator. Its energy is $E = p^2/(2m) + Kx^2/2$. In thermal contact with a reservoir at temperature T, its thermal energy is $k_B T$, since there are two degrees of freedom.

Now we come back to the value of Q. The expression (2.34) is the one-particle partition function of the ideal gas (in three dimensions). Below in the Chapter 4, we shall calculate this partition function from the quantum mechanical principles. In order to make the two results identical (in this chapter and in Chapter 4) one has to chose $Q = h$, the Planck constant, the Planck constant has effectively the dimension of (energy)(time).

2.2 The Grand Canonical Ensemble

In the present case, one supposes that the system and the reservoir are made by the same type of particles. The system is in thermal contact with the reservoir, but now the walls, which delimitate the volume V of the system, are permeable to the particles. The state of the system is determined by the temperature, volume and chemical potential of the reservoir. The energy and density are fluctuating quantities and one looks for their mean values that we consider equal to the macroscopic values.

2.2.1 *The grand partition function*

As above, we consider a large ensemble of identical systems. We are looking for the probability p_s to find the system in a microstate s with energy E_S and number of particles n_s. We begin with the same expressions as above by writing p_s as given in (2.1), but in specifying that the number of microstates are functions of the energy and the number of particles:

$$p_s = \frac{\Omega_R(E_0 - E_S, N_0 - n_s)}{\Omega_{GS}(E_0, N_0)}. \tag{2.36}$$

Let N_0 is the number of particles in the grand system, a constant, and $N_0 - n_s$ the number of particles in the reservoir. We write again as above

$$S_{GS} = k_B \operatorname{Ln} \Omega_{GS}(E_0, N_0),$$

$$S_R = k_B \operatorname{Ln} \Omega_R(E_0 - E_S, N_0 - n_s),$$

or

$$\Omega_{GS}(E_0) = \exp\left[\frac{S_{GS}(E_0, N_0)}{k_B}\right], \tag{2.37}$$

$$\Omega_R(E_0 - E_S) = \exp\left[\frac{S_R(E_0 - E_S, N_0 - n_s)}{k_B}\right]. \tag{2.38}$$

In the following steps we use the thermodynamic relation $E = TS - PV + \mu N$ to write S in the following form

$$S = \frac{PV}{T} + \frac{E - \mu N}{T}. \tag{2.39}$$

Substituting into (2.37) and (2.38) one gets

$$\Omega_{GS}(E_0, N_o) = \exp\left[\frac{P_{GS}V_{GS}}{T} + \frac{E_0 - \mu N_0}{k_B T}\right]$$

$$= \exp\left[\frac{P_{GS}V_{GS}}{k_B T}\right] \exp\left[\frac{E_0 - \mu N_0}{k_B T}\right]$$

$$= A_{GS} \exp\left[\frac{E_0 - \mu N_0}{k_B T}\right], \tag{2.40}$$

with $A_{GS} = \exp[(P_{GS}V_{GS})/(k_B T)]$.

For $\Omega_R(E_0 - E_s, N_0 - n_s)$, one has

$$\Omega_R(E_0 - E_s, N - n_s) = \exp\left[\frac{P_R V_R}{k_B T}\right] \exp\left[\frac{E_0 - E_s}{k_B T} - \frac{\mu(N_0 - n_s)}{k_B T}\right]$$

$$= A_R \exp\left[\frac{E_0 - E_s}{k_B T} - \frac{\mu(N_a - n_s)}{k_B T}\right]$$

$$= A_R \exp\left[\frac{E_0 - \mu N_0}{k_B T}\right] \exp\left[-\frac{E_S - \mu n_s}{k_B T}\right], \tag{2.41}$$

with $A_R = \exp[(P_R V_R)/(k_B T)]$.

Putting (2.40) and (2.41) in (2.36) gives

$$p_s = \frac{A_R}{A_{GS}} \exp\left[-\frac{E_S - \mu n_s}{k_B T}\right]. \tag{2.42}$$

The sum of the probabilities over all the microstates $\sum_S p_s = 1$. Thus

$$\sum_S p_s = \frac{A_R}{A_{GS}} \sum_S \exp\left[-\frac{E_s - \mu n_s}{k_B T}\right] = 1,$$

and this means that

$$\frac{A_{GS}}{A_R} = \sum_S \exp\left[-\frac{E_s - \mu n_s}{k_B T}\right].$$

Finally we arrive at the final result for p_s,

$$p_s = \frac{\exp\left[-\frac{E_s - \mu n_s}{k_B T}\right]}{\sum_s \exp\left[-\frac{E_s - \mu n_s}{k_B T}\right]}. \tag{2.43}$$

We stress again that the sum is over all the possible microstates of the system with all the possible energies and number of particles. The values of the mean energy and the mean number of particles are $E = \sum_S p_s E_S$ and $N = \sum_S p_s n_s$. But we can get these values through the grand partition function Z_G defined by

$$Z_G = \sum_s \exp[-\beta(E_S - \mu n_s)], \qquad (2.44)$$

with $\beta = 1/(k_B T)$. The sum (2.44) is in fact a double sum, on the microstates and on the number of particles. Z_G is a function of T, V and μ.

2.2.2   *The number of particles, energy, entropy and the grand potential*

The mean number of the particles N in the system is given by

$$N = \sum_s n_s p_S = \frac{\sum_s n_s \exp[-\beta(E_S - \mu n_s)]}{\sum_s \exp[-\beta(E_S - \mu n_s)]}. \qquad (2.45)$$

Recalling that that the numerator of (2.45) is the derivative of the denominator, relatively to μ, divided by β, and with a change in the sign, (2.45) can be written as

$$N = \frac{\frac{1}{\beta}\frac{\partial Z_G}{\partial \mu}}{Z_G} = \frac{1}{\beta}\frac{\partial \operatorname{Ln} Z_G}{\partial \mu}. \qquad (2.46)$$

The energy $E = \sum_s p_s E_S$ can be calculated with the help of the derivative $\frac{\partial \operatorname{Ln} Z_G}{\partial \beta}$, which is

$$\frac{\partial \operatorname{Ln} Z_G}{\partial \beta} = -\frac{\sum_s (E_S - \mu n_s)\exp[-\beta(E_S - \mu n_s)]}{\sum_s \exp[-\beta(E_S - \mu n_s)]}. \qquad (2.47)$$

The numerator of (2.47) is

$$-\sum_s E_S \exp[-\beta(E_s - \mu n_s)] + \sum_s \mu n_s \exp[-\beta(E_s - \mu n_s)],$$

and consequently $\partial \operatorname{Ln} Z_G / \partial \beta$ is the sum of two terms. The first is

$$-\frac{\sum_s E_S \exp[-\beta(E_S - \mu n_s)]}{\sum_s \exp[-\beta(E_S - \mu n_s)]}, \qquad (2.48)$$

which is $-\sum_s p_s E_S = -E$. The second term is

$$\frac{\sum_s \mu n_s \exp[-\beta(E_s - \mu n_s)]}{\sum_s \exp[-\beta(E_S - \mu n_s)]}$$

which is equal to

$$\mu \frac{\sum_s n_s \exp[-\beta(E_s - \mu n_s)]}{\sum_s n_s \exp[-\beta(E_S - \mu n_s)}$$

$$= \mu \sum_s p_s n_s = \mu N. \tag{2.49}$$

Finally one gets $\partial \ln Z_G / \partial \beta = -E + \mu N$, and for E one has

$$E = -\frac{\partial \ln Z_G}{\partial \beta} + \mu N = -\frac{\partial \ln Z_G}{\partial \beta} + \frac{\mu}{\beta} \frac{\partial \ln Z_G}{\partial \mu}, \tag{2.50}$$

when N was replaced by its value in (2.46), $(1/\beta)(\partial \ln Z_G / \partial \mu)$.

Since E is equal to $-\partial \ln Z / \partial \beta$, we have the following relation between the grand partition function and the partition function Z of a system with the same temperature, the same mean energy and a number of particles equal to the mean number of particles of our present system,

$$\frac{\partial \ln Z_G}{\partial \beta} = \frac{\partial \ln Z}{\partial \beta} + \mu N. \tag{2.51}$$

The entropy is determined using (2.51). We saw above, in the section on the canonical ensemble expression (2.16),

$$S = \frac{E}{T} + k_B \ln Z.$$

Integrating (2.51) gives

$$\ln Z_G = \ln Z - \int \mu N \, d\beta = \ln Z - \mu N \beta. \tag{2.52}$$

From (2.16) one gets

$$\ln Z = \frac{S - E/T}{k_B}. \tag{2.53}$$

Introducing this expression of $\ln Z$ in (2.52) gives

$$S = \frac{E - \mu N}{T} + k_B \ln Z_G. \tag{2.53}$$

From (2.50), $E - \mu N = -\partial \ln Z_G / \partial \beta$, and upon introducing it in (2.53),

$$S = \frac{-\frac{\partial \ln Z_G}{\partial \beta}}{T} + k_B \ln Z_G,$$

$$S = -k_B \beta \frac{\partial \ln Z_G}{\partial \beta} + k_B \ln Z_G. \tag{2.54}$$

It is not difficult to verify the relation $S = -k_B \sum_s p_s \ln p_s$, in analogy with the established cases.

From the preceding results, one deduces the grand potential Ψ given by

$$\Psi(T, V, N) = -PV = E - TS - \mu N \tag{2.55a}$$

$$= -k_B T \ln Z_G, \tag{2.55b}$$

where its differential is $d\Psi = -S\,dT + P\,dV - N\,d\mu$. To get (2.55b), one inserts in (2.55a) the value of $E - \mu N = \frac{\partial \ln Z_G}{\partial \beta}$ (see (2.50)), and S from (2.54). The expression (2.55) is the third link between the macroscopic thermodynamic description and the microscopic side.

2.2.3 *An example*

One considers a system in contact with a reservoir in the conditions of the grand canonical ensemble. This system is made of particles without interaction with energies $0, e, 2e, 3e,$ *etc.* The grand partition function is the double sum

$$Z_G = \sum_s \exp[-\beta(E_S - \mu n_s)],$$

which can be written as

$$Z_G = \sum_s [\exp(\beta \mu n_s) \sum_s \exp(-\beta E_S)]. \tag{2.56}$$

First one performs the sum over the microstates of a system with n_s particles and then to sum over the number of particles, from 0 to the infinity. The sum $\sum_s \exp(-\beta E_S)$ is the partition function of n_s particles (with the same volume V and temperature T). It is not possible to use the expression (2.21), $Z = (Z_1)^N$, since now the particles are indistinguishable. In the case of the grand canonical ensemble, the particles may leave the system to enter the reservoir, and vice versa, as a gas. This prevents the case when they are viewed as distinguishable. This fact makes the determination of the partition function difficult.

We shall consider a particular situation that we shall call the *classical limit* which in the next chapter we shall explain why. We suppose that particles are distributed among their possible energies or levels as follows: either one particular level is populated by one particle or it is not, i.e. there is no particle with this energy. When one considers a given repartition of the n_s particles in the energy levels, a particular microstate is defined. If now one changes the position of some particles, keeping the same occupied levels, one has the same microstate. To take into account this fact one divides the partition function by the number of possible permutations of the particles between the occupied levels, i.e. by $n_s!$ Thus the partition function of the n_s particles is

$$Z = \frac{(Z_1)^{ns}}{n_s!}. \tag{2.57}$$

We shall find again this expression by another method in Chapter 3.

The grand partition function is now

$$Z_G = \sum_s \left\{ \exp(\beta\mu n_s) \frac{[\sum_i \exp(-\beta e_i)]^{ns}}{n_s!} \right\}, \tag{2.58}$$

with $e_i = 0, e, 2e, 3e, etc.$

The sum $\sum_i \exp(-\beta e_i)$ is equal to $[1 - \exp(-\beta e)]^{-1}$, this gives

$$Z_G = \sum_s \frac{\exp(\beta\mu n_s)[1 - \exp(-\beta e)]^{-ns}}{n_s!} \tag{2.59}$$

$$= \sum_s \frac{\{\exp(\beta\mu)[1 - \exp(-\beta e)]^{-1}\}^{ns}}{n_s!}, \tag{2.60}$$

or with $x = \exp(\beta\mu)[1 - \exp(-\beta e)]^{-1}$,

$$Z_G = \sum_s \frac{x^{ns}}{n_s!} = \exp(x), \tag{2.61}$$

where we used the series expansion of the function $\exp(x)$. Our final result is

$$Z_G = \exp\{\exp(\beta\mu)[1 - \exp(-\beta e)]^{-1}\}, \tag{2.62}$$

$$\text{Ln}\, Z_G = \exp(\beta\mu)[1 - \exp(-\beta e)]^{-1}, \tag{2.63}$$

$$\Psi = -k_B T \,\text{Ln}\, Z_G = -k_B T \exp(\beta\mu)[1 - \exp(-\beta e)]^{-1}. \tag{2.64}$$

Now one can calculate the mean number of particles N

$$N = \frac{1}{\beta}\frac{\partial \text{Ln}\, Z_G}{\partial \mu} = \exp(\beta\mu)[1 - \exp(-\beta e)]^{-1}. \tag{2.65}$$

Putting this expression in (2.64) gives the equation of state

$$-\Psi = PV = N k_B T. \tag{2.66}$$

From (2.65) one gets an expression for the chemical potential μ

$$\mu = k_B T[\text{Ln}(N) + \text{Ln}(1 - \exp(-\beta e))]. \tag{2.67}$$

The derivative of $\text{Ln}\, Z_G$ relatively to β is, following (2.50),

$$\frac{\partial \text{Ln}\, Z_G}{\partial \beta} = \mu N - E. \tag{2.68}$$

One has

$$\frac{\partial \text{Ln}\, Z_G}{\partial \beta} = \mu \exp(\beta\mu)[1-\exp(-\beta e)] - \frac{\exp(\beta\mu)\, e \exp(-\beta e)}{[1 - \exp(-\beta e)]^2} \tag{2.69}$$

Taking into account of (2.65) one can write

$$\frac{\partial \text{Ln}\, Z_G}{\partial \beta} = \mu N - Ne\frac{\exp(-\beta e)}{1 - \exp(-\beta e)} \tag{2.70}$$

Comparing (2.68) and (2.70) gives

$$E = Ne\frac{\exp(-\beta e)}{1 - \exp(-\beta e)} = \frac{Ne}{\exp(\beta e) - 1}. \tag{2.71}$$

The entropy can be calculated through the formula $E = TS - PV + \mu N$, giving

$$S = \frac{E}{T} + \frac{PV}{T} - \frac{\mu N}{T}.$$

Taking into consideration (2.66), (2.67) and (2.71) one gets

$$\frac{S}{k_b N} = \frac{\frac{e}{k_B T}}{\exp(e/k_B T) - 1} + 1 - \text{Ln}\, N - \text{Ln}\left[1 - \exp\left(-\frac{e}{k_B T}\right)\right]. \tag{2.72}$$

In the next chapter we shall show that, in the conditions we choose, the chemical potential is negative such that the entropy is always positive. It is also possible to verify that S is an increasing function of the temperature.

The condition that $\mu < 0$ gives the regime of validity of this problem. Writing, from (2.67), that

$$\frac{\mu}{k_B T} = \text{Ln } N + \text{Ln}(1 - \exp(-\beta e)) < 0$$

gives

$$\text{Ln } N + \text{Ln}[1 - \exp(-\beta e)] = \text{Ln}[N(1 - \exp(-\beta e))] < 0$$

or

$$N[1 - \exp(-\beta e)] < 1.$$

Since N is very large, this means that $1 - \exp(-\beta e)$ is very small. In other words, $\exp(-\beta e)$ is very close to 1, i.e. βe is very small. Using the approximation $\exp(-\beta e) \approx 1 - \beta e$, the inequality $N[1 - \exp(-\beta e)] < 1$ becomes $\beta e < 1/N$ or $T > eN/k_B$ .This means that the temperature is large enough and/or e (the distance between two consecutive levels) is very small. We shall see later that this situation is called the classical case.

2.3 Summary

We summarize the developments of the two last chapters. We exposed three methods, which establish the links between the microscopic description and the thermodynamics or macroscopic description of a system.

The microcanonical ensemble. The system is defined by the knowledge of the extensive variables like the energy E, the volume V, the number of particles N, etc. The thermodynamic potential is the entropy and the link is made by the relation

$$S(E, V, N) = k_B \text{ Ln } \Omega,$$

where Ω is the number of microstates of the system associated with the chosen values of the extensive variables. In general it is not an easy task to find $\Omega(E, V, N)$ and it is why the method of the microcanonical is not frequently used.

The canonical ensemble. The system is defined through knowledge of the temperature, the volume and the number of particles. The thermodynamic potential is the Helmholtz free energy F and the link is the partition function Z:

$$F(T, V, N) = -k_B T \operatorname{Ln} Z$$

with Z equal to $\sum_s \exp(-\beta E_S)$. The sum is over all the microstates of the system. In this method, we need to determine the energies of all the microstates of the system. In the following, we shall use very often this method.

The grand canonical ensemble. In this case, the temperature, the volume and the chemical potential define the system. The thermodynamic potential is the grand potential Ψ and the link is the grand partition function Z_G:

$$\Psi(T, V, \mu) = -k_B T \operatorname{Ln} Z_G$$

with $Z_G = \sum_s \exp[-\beta(E_S - \mu n_s)]$. The sum is a double sum over the possible energies of the microstates and the number of particles. This method seems complicated but it permits the introduction of a very important variable: the chemical potential. We shall see in the following chapters the usefulness of this variable.

In the last two methods, the goal is to calculate mean values of the variables that we take as the macroscopic values, neglecting fluctuations. This is possible, in general, because we deal with a very large number of particles. But near a critical point or a phase transition of second kind, this is not correct and the fluctuations need to be taken into account.

Nevertheless, the expressions that we got for p_s the probability to find the system in a given microstate s is correct whatever the number of particles is, even for one particle in the case of the canonical ensemble. Of course, in such a case, the fluctuations are important.

The entropy. The most general expression for the entropy, in these three approaches is

$$S = -k_B \sum_s p_s \operatorname{Ln} p_s$$

where p_s is the probability to find the system in the microstate s.

2.3.1 *Fluctuations*

We have supposed that the mean value of the energy (and other quantities) is equal to its macroscopic value although we know that this quantity fluctuates in the case of the canonical and grand canonical ensembles.

In the case of the thermal contact of the system with the reservoir, it was stated that the temperature is fixed when the energy is fluctuating. This means that there is constant transfer of energy from the reservoir to the system and vice versa. How is this possible if there is no temperature difference between the system and the reservoir? One possibility is the transfer of potential energy through the variations of the distances between particles since this kind of energy is dependant on distances. But in fact there are changes in the temperatures although they are so small that one does not consider them and accepts to say that the temperature is constant.

In thermodynamics, it is possible to calculate the mean standard deviations of the temperature and the energy in the system. It is the mean value of the squared difference between the fluctuating quantity and its mean value. We give the results without derivations. For the temperature one has $(\Delta T)^2 = k_B T^2/C_v$ where C_v is the constant volume specific heat, and for the energy one has $(\Delta E)^2 = k_B T^2 C_v$. The dimensionless ratio $\delta = k_B[(\Delta T)^2/(\Delta E)^2]^{0.5}$ can give a feeling of the relative magnitude of the fluctuations. One finds that $\delta = k_B/C_v$. We take as an example the monatomic ideal gas. In this case C_v is equal to $(3/2)\,N\,k_B$ and the ratio δ is equal to $2/(3N)$. For a mole, N is of order of 10^{23} and δ is very small. The fluctuation of the temperature is much smaller than those of the energy. The conclusion is that minuscule changes in the temperature may bring important changes in the energy.

2.3.2   *Final remark*

One can point out the analogy between the formulas giving the three thermodynamic potentials, $-S$, F and Ψ:

$$-S = -k_B \operatorname{Ln} \Omega;$$

$$F = -k_B T \operatorname{Ln} Z;$$

$$\Psi = -k_B T \operatorname{Ln} Z_G.$$

Chapter 3

Quantum Statistics

In this chapter, we shall calculate the general expression for the free energy from the partition function or the grand partition function of a gas, when quantum effects are present. We shall see later that in certain conditions the quantum effect is not important and we shall call the gas in such circumstances a classical gas or an ideal gas. As in the rest of this book, we consider the simple case of particles without interaction.

In quantum mechanics, one distinguishes between two kinds of particles: the bosons and the fermions. There are two basic differences between them. The first difference stands in the spin. The spin of a particle is $\mathrm{nh}/(2\pi)$ when n takes integer values (0, 1, 2, *etc.*) for the bosons and half-integer values (1/2, 3/2, 5/2, *etc.*) for the fermions. This means that a state of a particle is also characterized by the z components s_z of the spin which can take values between $-n$ and $+n$, by steps of one $(-n, -n+1, -n+2, \ldots, n-1, n)$. Thus, the state of an isolated particle is defined by the vector linear momentum **p** and the value of s_z.

The second difference is that the fermions obey the Pauli principle: two fermions in a particular group of fermions cannot be in the same state when for the bosons such restriction does not exist. Two or more bosons can be in the same state. The fermions obey the Fermi–Dirac statistics and the bosons the Bose–Einstein statistics.

44

Examples of fermions are the electron, the proton and the neutron; examples of bosons are the photon and the α particle.

3.1 The Partition Function and the Free Energy

When one considers a quantum gas, two different situations can happen. The number of particles may or may be not fixed. In the first case, we indicate the number of particles by N, and in the second case this number can fluctuate. One can only speak about the mean number of particles, which is, in general, dependent on the temperature. A well known example of a gas with a non-fixed number of particles is a gas of photons.

An isolated particle has a number of possible states that we label by a series of numbers 1, 2, 3, and so on. To each state i, corresponds an energy e_i, and it is possible that two or more different states have the same energy. Since the particles are confined in a volume V, the possible energy e_i form a discontinuous series of values. We suppose that all the e_i's are positive. A microstate of the system of particles is defined by the number n_1 of particles in the state 1, n_2 particles in the state 2, ..., n_i particles in the state i, *etc.*. The ensemble $\{n_i\}$ of the number of particles in each state is characteristic of a microstate. If one permutes two particles between the two states to which they belong, one has the same microstate, since in quantum mechanics it is not possible to distinguish between two particles.

The energy of the system for a given microstate, i.e. for a given ensemble of the n_i's is

$$E_S = \sum_{ni} n_i e_i \tag{3.1}$$

with the following conditions: a) For fermions, n_i can be equal to zero or to one (impossibility to have two particles in the same state), and for bosons n_i can take all the possible values; b) If the number of particles is fixed, one has $N = \sum n_i$.

The partition function is

$$Z = \sum \exp[-\beta(\sum n_i e_i)], \tag{3.2a}$$

or

$$Z = \sum \exp[-\beta(n_1 e_1 + n_2 e_2 + \cdots + n_i e_i + \cdots)], \tag{3.2b}$$

when the sums are over all the microstates or over all the possible ensembles $\{n_i\}$. We write again (3.2) in a different form. Putting $z_i = \exp(-\beta e_i) < 1$, one has

$$Z = \sum (z_1)^{n1}(z_2)^{n2}(z_3)^{n3} \cdots (z_i)^{ni} \cdots \qquad (3.3)$$

3.1.1 *For variable N and for fermions*

In the sum (3.3), n_i can be equal to zero or to one, and the sum (3.3) is without limit about the total number of particles. We write (3.3), summing first on n_1

$$Z = (z_1)^0 \sum (z_2)^{n2}(z_3)^{n3}(z_4)^{n4} \cdots$$
$$+ (z_1) \sum (z_2)^{n2}(z_3)^{n3}(z_4)^{n4} \cdots \qquad (3.4)$$

or

$$Z = (1 + z_1) \sum (z_2)^{n2}(z_3)^{n3}(z_4)^{n4} \cdots . \qquad (3.5)$$

One can repeat this process for z_2 and then for z_3 and so on. We get

$$Z = (1 + z_1)(1 + z_2) \cdots (1 + z_i) \cdots , \qquad (3.6a)$$
$$Z = \prod_i (1 + z_i). \qquad (3.6b)$$

Now the free energy F is equal to $-k_B T \operatorname{Ln} Z$ and also to

$$F = -k_B T \operatorname{Ln} \left[\prod_i (1 + z_i) \right], \qquad (3.7a)$$

or

$$F = -k_B T \sum_i \operatorname{Ln}[1 + \exp(-\beta e_i)]. \qquad (3.7b)$$

3.1.2 *For variable N and for bosons*

We used the same procedure as done for the fermions, taking into account that there is no limitation on the n_i's. We begin by summing on the possible values of n_1:

$$Z = (z_1)^0 \sum (z_2)^{n2}(z_3)^{n3}(z_4)^{n4} + \cdots + (z_1) \sum (z_2)^{n2}(z_3)^{n3}(z_4)^{n4} \cdots$$
$$+ (z_1)^2 \sum (z_2)^{n2}(z_3)^{n3}(z_4)^{n4} + \cdots$$
$$+ (z_1)^3 \sum (z_2)^{n2}(z_3)^{n3}(z_4)^{n4} \cdots ,$$

(3.8)

or

$$Z = [1 + z_1 + (z_1)^2 + (z_1)^3 + \cdots] \sum (z_2)^{n2}(z_3)^{n3}(z_4)^{n4} \cdots . \quad (3.9)$$

Since there is no limit for the possible values of the number n_1, the sum in the bracket is infinite. Since $z_i < 1$ (we recall that $e_i > 0$), the series has a limit, $(1 - z_1)^{-1}$, and one has

$$Z = (1 - z_1)^{-1} \sum (z_2)^{n2}(z_3)^{n3}(z_4)^{n4} \cdots . \quad (3.10)$$

As above, we repeat the process for n_2, then for n_3 and so on. We obtain

$$Z = (1 - z_1)^{-1}(1 - z_2)^{-1}(1 - z_3)^{-1} \cdots , \quad (3.11)$$

$$Z = \Pi_i (1 - z_i)^{-1}. \quad (3.12)$$

Finally for the free energy, one has

$$F = -k_B T \sum_I -\mathrm{Ln}[1 - \exp(-\beta e_i)]. \quad (3.13)$$

When one compares the expressions for the fermions (3.6) and (3.7) with those for the bosons (3.12) and (3.13), one sees that the difference stands in a sign $+$ for the fermions and $-$ for the bosons. This permits to write in a compact form the free energy in the case of non-fixed N:

$$F = -k_B T \sum_i \pm \mathrm{Ln}[1 \pm \exp(-\beta e_i)], \quad (3.14)$$

and

$$Z = \Pi_i (1 \pm z_i)^{\pm 1}, \quad (3.15)$$

when one puts the sign $+$ for the fermions and the sign $-$ for the bosons.

3.1.3,4 *For fixed N, and for both fermions and bosons*

We deal with both cases for fermions and bosons together since, as we shall show below, the calculations are very similar to the precedent cases.

The calculations of the sum (3.3) is difficult since we have to take into account only the possible ensembles of $\{n_i\}$ for which the total number of particles is fixed, $\sum_i n_i = N$. We avoid this difficulty in calculating the grand partition function. But there is a price to pay, we have to introduce the chemical potential and the results are more complicated. The grand partition function Z_G is given by

$$Z_G = \sum_s \exp[-\beta(E_s - \mu n_s)],$$

where the sum is a double sum, on the possible states of the system and the number of particles. Z_G can be written as

$$Z_G = \sum_s \exp\{-\beta[\sum_i (n_i e_i) - \mu n_s]\}. \tag{3.16}$$

To calculate this sum, we proceed, according to principle, by steps: first we choose a value of n_s, distribute these n_s particles in the different states such that $\sum_i n_i = n_s$, and perform the sum. In a second step, we choose a different value of n_s and repeat the procedure. We can operate differently by writing Z_G in the following form

$$Z_G = \sum_s \exp[-\beta(\sum_i n_i e_i - \mu \sum_i n_i)]$$

$$= \sum_{ni} \exp[-\beta \sum_i n_i(e_i - \mu)], \tag{3.17}$$

or putting $z_i = \exp[-\beta(e_i - \mu)]$,

$$Z_G = \sum_{ni} (z_1)^{n1}(z_2)^{n2}(z_3)^{n3} \cdots (z_i)^{ni} \cdots, \tag{3.18}$$

which has exactly the same form as (3.3). The calculations are identical to those we did above, since in the grand canonical ensemble we perform the sum without any limitation on the number of particles. We can write immediately the results from the precedent section:

$$Z_G = \Pi_i(1 \pm z_i)^{\pm 1} \tag{3.19}$$

with $z_i = \exp[-\beta(e_i - \mu)]$ and, as above, the sign $+$ is for the fermions and the sign $-$ for the bosons. From (3.19), one has

$$\text{Ln}\, Z_G = \sum_I \text{Ln}[(1 \pm z_i)^{\pm 1}]$$

$$= \sum_i \pm \text{Ln}\{1 \pm \exp[-\beta(e_i - \mu)]\}. \tag{3.20}$$

Finally, we deduce from the expression (2.40), $E - TS - \mu N = -k_B T \operatorname{Ln} Z_G$, the free energy $F = E - TS = -k_B T \operatorname{Ln} Z_G + \mu N$, or

$$F = -k_B T \sum_i \pm \operatorname{Ln}\{1 \pm \exp[-\beta(e_i - \mu)]\} + \mu N. \qquad (3.21)$$

N is given by

$$N = \frac{1}{\beta} \frac{\partial \operatorname{Ln} Z_G}{\partial \mu} = \sum_i \exp[-\beta(e_i - \mu)]\{1 \pm \exp[-\beta(e_i - \mu)]\}^{-1},$$

$$(3.22a)$$

or by multiplying the numerator and the denominator by $\exp[\beta(e_i - \mu)]$

$$N = \sum_i \{\exp[\beta(e_i - \mu)] \pm 1\}^{-1} \qquad (3.22b)$$

The expression (3.22b) gives the possibility to calculate the chemical potential $\mu(T, V, N)$ and from the result one can calculate $F(T, V, N)$.

When one compares the expression (3.20) with (3.14) giving the free energy for the case of non-fixed N, one sees that (3.14) is a particular case of (3.21) with $\mu = 0$. Conclusion: the expressions (3.20) and (3.21) are always valid with the condition $\mu = 0$ for non-fixed N.

3.2 The Energy and the Entropy

The energy is given by the expression (2.37), $E = -\partial \operatorname{Ln} Z_G / \partial \beta + (\mu/\beta)(\partial \operatorname{Ln} Z_G / \partial \mu)$ or $E = -\partial \operatorname{Ln} Z_G / \partial \beta) + \mu N$. One has

$$(\partial \operatorname{Ln} Z_G / \partial \beta) = \sum_i -(e_i - \mu) \exp[-\beta(e_i - \mu)]$$

$$\times [1 \pm \exp[-\beta(e_i - \mu)]]^{-1}$$

$$= \sum_i -e_i \exp[-\beta(e_i - \mu)][1 \pm \exp[-\beta(e_i - \mu)]]^{-1}$$

$$+ \mu \sum_i \exp[-\beta(e_i - \mu)][1 \pm \exp[-\beta(e_i - \mu)]]^{-1},$$

and using (3.22a),

$$-\frac{\partial \operatorname{Ln} Z_G}{\partial \beta} = \sum_i e_i \exp[-\beta(e_i - \mu)][1 \pm \exp[-\beta(e_i - \mu)]]^{-1} - \mu N,$$

and finally

$$E = \sum_i e_i \exp[-\beta(e_i - \mu)][1 \pm \exp[-\beta(e_i - \mu)]]^{-1}$$

Multiplying the numerator and the denominator by $\exp[\beta(e_i - \mu)]$, one gets

$$E = \sum_i e_i \{\exp[\beta(e_i - \mu)] \pm 1\}^{-1}. \qquad (3.23)$$

This can be written again as $E = \sum_r n_r e_r$ where n_r is the average number of particles in the state r with energy e_r (we changed the labeling from n_i, the number of particles in state i at a given time, to n_r, the mean number of particles in state r for a given macrostate with energy E). One has

$$n_r = \{\exp[\beta(e_r - \mu)] \pm 1\}^{-1}. \qquad (3.24)$$

This expression is compatible with (3.22b) since it can be read as $N = \sum_r n_r$. In the case of the fermions, the n_i's can be equal to zero or to one, and as shown by (3.24) the mean number is always smaller than one. But in the case of bosons, since n_r must be a positive number, one has $\exp[\beta(e_r - \mu)] > 1$ for all the possible values of e_i. This is possible only if μ is smaller than the lowest value of the e_i's. In particular, if this value is chosen to be zero, the chemical potential of a group of bosons is negative except for the case of non-fixed particles when it is zero.

The expressions (3.23) and (3.24)(N is the mean number) are also valid for particles with a non-fixed number; it suffices to make $\mu = 0$.

To calculate the entropy, we use result of the preceding chapter, expression (2.39), $S = (E - \mu N)/T + k_B \operatorname{Ln} Z_G$. We have

$$S = \frac{E - \mu N}{T} + k_B \sum_i \pm \operatorname{Ln}[1 \pm \exp[-\beta(e_i - \mu)]]. \qquad (3.25)$$

3.3 The Classical Ideal Gas: Maxwell–Boltzmann Statistics

By definition, a classical gas is a gas in which the quantum effects are negligible. This situation may happen if either $\exp[\beta(e_i - \mu)] \gg 1$

or $\exp[-\beta(e_i - \mu)] \ll 1$, such that the expressions for the bosons and those for the fermions become identical. Consequently $\exp[\beta(e_i - \mu)] \pm 1$ is practically equal to $\exp[\beta(e_i - \mu)]$ and there is no difference between fermions or bosons. Since

$$n_i = \exp[-\beta(e_i - \mu)] \ll 1,$$

this situation is possible if practically every level is either occupied by only one particle or not occupied at all. Clearly in this case the mean number of particles in a given level is very small.

This condition must be true even for the lowest e_i which we take equal to zero. This means that $\exp(-\beta\mu) \gg 1$ or that μ is strongly negative (as we have supposed in the third example of the preceding chapter). Such situation occurs at high temperature, associated with the property $(\partial\mu/\partial T)_{V,N} < 0$. We postpone the demonstration of this statement to the next section. In the case of the gas with a non-fixed number of particles, we cannot reach such a condition (since $\mu \equiv 0$); a classical gas has always a well defined number of particles.

The condition $\exp(-\beta\mu) \gg 1$, or $\exp(\beta\mu) \ll 1$, gives the following results for the mean number of particles in the state i

$$n_i = \exp(\beta\mu)\exp(-\beta e_i), \qquad (3.26a)$$

and for the total number N, by (3.22b),

$$N = \exp(\beta\mu)\sum_i \exp(-\beta e_i). \qquad (3.26b)$$

The sum $\sum_i \exp(-\beta e_i)$ is the partition function Z_1 of a particle, and one has

$$N = \exp(\beta\mu)\,Z_1. \qquad (3.26c)$$

From (3.26a) and (3.26b) one gets the important result of the classical case

$$\frac{n_i}{N} = \frac{\exp(-\beta e_i)}{Z_1}. \qquad (3.27)$$

The chemical potential is obtained from the expression (3.26c) of N

$$\beta\mu = -\mathrm{Ln}\!\left(\frac{Z_1}{N}\right), \qquad (3.28)$$

or

$$\mu = -k_B T \operatorname{Ln} Z_1 + k_B T \operatorname{Ln} N. \tag{3.29}$$

We calculate now the Helmholtz free energy F. Since $\exp[\beta(e_i - \mu)] \gg 1$, or $\exp[-\beta(e_i - \mu)] \ll 1$, one can use, in the expression of F, (3.21), the approximation $\operatorname{Ln}(1 + x) \approx x$ for $x \ll 1$, and one obtains

$$F = -k_B T \sum_i \exp[-\beta(e_i - \mu)] + \mu N, \tag{3.30a}$$

$$F = -k_B T \exp(\beta\mu) \sum_i \exp(-\beta e_i) + \mu N. \tag{3.30b}$$

One inserts in (3.30b) the expressions (3.27) for $\exp(\beta\mu)$ and (3.29) for μ, and taking into account that $Z_1 = \sum_i \exp(-\beta e_i)$ is the partition function of a particle,

$$F = -k_B T \left(\frac{N}{Z_1}\right) Z_1 + N(-k_B T \operatorname{Ln} Z_1 + k_B T \operatorname{Ln} N),$$

or

$$F = -k_B T (N \operatorname{Ln} Z_1 + N - N \operatorname{Ln} N). \tag{3.31}$$

If N is very large, as we suppose, we can use the Stirling formula $\operatorname{Ln} N! \cong N \operatorname{Ln} N - N$, and we write F as

$$F = -k_B T [\operatorname{Ln}(Z_1)^N - \operatorname{Ln} N!], \tag{3.32}$$

$$F = -k_B T \operatorname{Ln}\left[\frac{(Z_1)^N}{N!}\right]. \tag{3.33}$$

We conclude that the partition function of the classical gas is

$$Z = \frac{(Z_1)^N}{N!}. \tag{3.34}$$

We note that it is different from the expression (2.20) of the partition function of N independent particles. Here, the quantity $N!$ appears in the denominator. The difference is due to the fact that in the present case of the gas, it is not possible to distinguish between the particles, in contrary to the case of the expression (2.20). One has to divide by all the possible permutations of the particles.

To finish this section, we come back to the expression (3.29) on the chemical potential. If it is strongly negative (as necessary for the ideal gas), this means that $Z_1 \gg N$. Since Z_1 is a function of T and V, this condition can be translated in another condition concerning only the variables T, V and N. We will give explicitly this condition in the next chapter when we will calculate the partition function of the ideal gas. This condition can be also written as $n_r \ll 1$, that the mean number in a particular state is much smaller than 1.

It is conventional to call such a situation of a classical gas the Maxwell–Boltzmann statistics. In the present case, the particles are indistinguishable.

3.4 Qualitative Behavior of the Chemical Potential and the Derivation of $(\partial\mu/\partial T)_{V,N} < 0$

We begin by the expression (3.22b) for N,

$$N = \sum_i \{\exp[\beta(e_i - \mu)] \pm 1\}^{-1},$$

and derive both sides relatively to T. Since N is constant, $dN/dT = 0$. One writes

$$\frac{dN}{dT} = \frac{dN}{d\beta}\frac{d\beta}{dT} = \frac{d\beta}{dT} \sum_i \left[(e_i - \mu) - \beta\left(\frac{\partial\mu}{\partial\beta}\right)\right]\frac{\exp[\beta(e_i - \mu)]}{D} = 0,$$

when D is equal to $\{\exp[\beta(e_i - \mu)] \pm 1\}^2$. In the expression of dN/dT, $dT/d\beta \neq 0$ and $D \neq 0$, such that dN/dT can be equal to zero only if the sum $\sum_i [(e_i - \mu) - \beta(\partial\mu/\partial\beta)]\exp[\beta(e_i - \mu)]$ is equal to zero. We can decompose it into two equal sums:

$$\sum_i (e_i - \mu)\exp[\beta(e_i - \mu)] = \beta\left(\frac{\partial\mu}{\partial\beta}\right)\sum_i \exp[\beta(e_i - \mu)], \quad (3.35)$$

or

$$\frac{\partial\mu}{\partial\beta} = \frac{\sum_i (e_i - \mu)\exp[\beta(e_i - \mu)]}{\beta\sum_i \exp[\beta(e_i - \mu)]}. \quad (3.36)$$

3.4.1 *Bosons*

In the case of bosons $e_i - \mu$ is positive since μ is smaller than all the e_i's. The numerator of (3.35) is positive as well as the denominator, since the sum $\sum_i \exp[\beta(e_i - \mu)]$ is positive. One concludes that $\partial\mu/\partial\beta > 0$. Since $\partial\mu/\partial T = (\partial\mu/\partial\beta)(d\beta/dT)$ and $d\beta/dT = -1/T^2 < 0$, one concludes that $\partial\mu/\partial T < 0$.

3.4.2 *Fermions*

The case of the fermions is a little bit more complicated. We suppose, as is above, without loss of generality that all the e_i's are positive and that the lowest is equal to zero. First, we note that for T going to zero, μ goes to a positive value. At $T = 0$, the system has the lowest possible energy. This means that the particles are in the lowest possible states. In the case of fermions, there is no more than one particle in each state, and consequently the states occupied by only one particle are the first N states. Recalling that the mean number in a state labeled r is $n_r = \{\exp[\beta(e_r - \mu)] + 1\}^{-1}$, one sees that, $\beta \to \infty$, i.e. $T \to 0$,

$$\text{if } e_r < \mu, \quad \exp[\beta(e_r - \mu)] \to 0 \text{ and } n_r \to 1;$$
$$\text{whereas if } e_r > \mu, \quad \exp[\beta(e_r - \mu)] \to \infty \text{ and } n_r \to 0.$$

Since $n_r = 1$ for the N first states, we conclude that μ is larger than or equal to the energy of the state with the largest occupied energy, i.e. the state $i = N$. Thus one gets the important result, $\mu(T = 0) > 0$.

Secondly, there is only one temperature $T_o = 1/k_B\beta_o$ for which $\mu = 0$. It is given by $N = \sum_i [\exp(\beta_o e_i) + 1]^{-1}$. Thus for $T < T_o$ one has $\mu > 0$, and consequently for $T > T_o$ one has $\mu < 0$ (because T_o is unique).[1]

Finally, for $\mu \leq 0$, $\partial\mu/\partial\beta > 0$ or $\partial\mu/dT < 0$, by applying the same reasoning that we used for the bosons. From those results, we deduce that for $T < T_o$, μ is positive, and that for $T > T_o$, μ is negative with a negative derivative $\partial\mu/\partial T < 0$. The case that μ may have a maximum in the region $\mu > 0$ is not excluded.

The conclusion is, for bosons the chemical potential is always negative, with a negative derivative relative to the temperature, when for fermions it is the case only for T exceeding than a particular temperature T_o.

---

[1] We shall see in the second part that it is effectively the case for fermions gas.

Chapter 4

The Density of States

In the definition of the partition and grand partition function, we insisted in specifying that the sums are over all the microstates of the system. In this chapter, we shall consider the case of a gas of particles without interaction and find the number $g_E(E)$ which gives, for a single particle, the number of states with energy E. The partition function is $Z = \sum_s \exp(-\beta E_s)$ with sum over all the states, and is $\sum_s g_E(E) \exp(-\beta E_s)$ with sum over the energies.

Since there is no interaction, the energy of a particle is the kinetic energy if the particle has finite mass or finite relativistic energy if the mass is zero. However, as we shall see later (in the next chapter), a molecule compounded of several atoms can have also internal energy. For the time, we deal with the kinetic energy which is equal to $E = p^2/(2m)$ when p is the value of the linear momentum and m is the mass of the particle. For particles with $m = 0$ like photons, the energy is relativistic, related to the momentum by $E = pc$ where c is the velocity of light.

The state of a particle is defined by the vector $\mathbf{p}$ and the z component s_z of the spin. It suffices to know what is the number of the possible values of s_z, and to multiply by the number of states with different vectors $\mathbf{p}$ but equal kinetic energy E, to get the function $g_E(E)$, called the density of states. We begin by finding out the number of states with the same value of p, i.e. with the same energy $E(p)$.

The second part of this chapter is devoted to the study of the ideal gas as an example of application of the density of states.

4.1 The Wave Vector

A particle in a box of dimensions L_x, L_y and L_z (volume $V = L_x L_y L_z$) is described in quantum mechanics by a wave function characterized by a wave vector $\mathbf{k}$ (with components k_x, k_y and k_z),

$$\Psi(x, y, z) = \Psi_0 \sin(k_x\, x) \sin(k_y\, y) \sin(k_z\, z) \qquad (4.1)$$

with $k_x = n_x \pi / L_x$, $k_y = n_y \pi / L_y$ and $k_z = n_z \pi / L_z$ where n_x, n_y and n_z are integers (0, 1, 2, 3, *etc.*). For the sake of simplicity, one chooses $L_x = L_y = L_z = L$. This choice does not change the final result while the calculations simplifying. One can write the square of the vector $\mathbf{k}$:

$$k^2 = (\pi^2/L^2)(n_x^2 + n_y^2 + n_z^2). \qquad (4.2)$$

The relation between the linear momentum and the wave vector is $\mathbf{p} = \hbar\mathbf{k}$, and the state of the particle is defined by the vector $\mathbf{p}$, or the vector $\mathbf{k}$, or by the three numbers n_x, n_y, n_z. One defines a number n as $n^2 = n_x^2 + n_y^2 + n_z^2$, and for each ensemble of states defined by three numbers n_x, n_y and n_z having the same n, the energy E has the same value. The possible values of k are discrete, with the exception of near zero temperatures, when the energy of the particles is such that the values of n are huge. For example, for an electron with energy of $1\,\mathrm{eV}$ located in a cubic box of sides $1\,\mathrm{mm}$, n is of the order of 10^6, and for a photon with the same energy in the same box, n is of the order of 10^{19}. This justifies taking the values of n as a continuum, since the two resultant values of n correspond to a minute difference in the energy. In this condition, the sum giving the partition function is replaced by an integral

$$Z = \int g_E(E) \exp(-\beta E)\, dE.$$

This can be also be written, with the help of p, as

$$Z = \int g_p(p) \exp[-\beta E(p)]\, dp.$$

In the case of quantum particles, one has, for $\text{Ln} \, Z_G$,

$$\text{Ln} \, Z_G = \int \pm g_E(E) \, \text{Ln} \, \{1 \pm \exp[-\beta(E - \mu)]\} \, dE.$$

In the expressions of Z and $\text{Ln} \, Z_G$ above, $g_E(E) \, dE$ is the number of states with energies between E and $E + dE$.

4.2 The Density of States

We have to count how many sets of the three numbers n_x, n_y and n_z there are located between n and $n + dn$. We indicate this number by $g_n(n) \, dn$. We considers a three-dimensional virtual space such that a point in this space is marked by three positive numbers on the three orthogonal axes n_x, n_y and n_z. In other words, we represent a state of the particle by a point in this space. Since the numbers n_x, n_y and n_z are positive, the points corresponding to all possible states of the particles are located only in $1/8$ of the total space. The number of points between the two shells of radius n and $n + dn$ are the number of states $g_n(n) \, dn$ that we have to find. It is

$$g_n(n) \, dn = \frac{1}{8} 4\pi n^2 \, dn. \tag{4.3}$$

($4\pi n^2$ is the surface of one of the two spheres delimiting the shell, and dn is the thickness of the shell.)

We recall the relation between k and n, $k = n(\pi/L)$, and that between k and p, $p = \hbar k = k(\hbar/2\pi)$, to get the relation between n and p: $n = p(2L/\hbar)$. Thus we get

$$g_n(n) \, dn = \frac{1}{8} 4\pi n^2 \, dn = \frac{\pi}{2} \left[p \frac{2L}{h} \right]^2 \left[\frac{2L}{h} dp \right], \tag{4.4a}$$

$$g_p(p) \, dp = \frac{4\pi L^3 p^2 \, dp}{h^3}. \tag{4.4b}$$

Finally, by taking into account the different states with different spin components, and that $V = L^3$, and since $g_n(n) \, dn = g_p(p) \, dp$, thus

$$g_p(p) \, dp = s_z \frac{(4\pi V p^2 \, dp)}{h^3} \tag{4.4c}$$

This is our final result which we shall use it frequently in the following chapters. The function $g_E(E)$ is easily deduced from the equality $g_p(p) \, dp = g_E(E) \, dE$ and the relation between p and E.

Important remark. Concerning the possible energies of a particle in a box. The numbers n_x, n_y and n_z appearing in (4.2) cannot be equal to zero together. In such a case, the wave function (4.1) is null, i.e. there is no particle in the box. The smallest values of the energy are those with one of n_x, n_y or n_z equal to 1 and the other two equal to zero (degeneracy 3 for the energy). If in the following chapters we shall take the energy zero as the lowest energy; it is done only as a change in the energy scale.

4.3 The Monatomic Ideal Gas

4.3.1 *The partition function*

The first step is to calculate the partition function $Z = Z_1^N/N!$ (expression (3.33) of the preceding chapter) when Z_1 is the partition function of an atom. Z_1 can be written as an integral (instead of a sum) since we have shown that, for an isolated particle in a box, the different possible energies form a continuum:

$$Z_1 = \int_0^\infty g_p(p) \exp\left[\frac{-p^2}{\frac{2m}{k_BT}}\right] dp = \frac{4\pi V}{h^3} \int_0^\infty p^2 \exp\left(\frac{-\beta p^2}{2m}\right) dp,$$

(4.5)

when (4.4) was used and we assumed that $s_z = 1$. The limits of the integration are from 0 to ∞.

In order to calculate the integral (4.5), we make a change of variables, and write $x = \beta p^2/(2m)$ and obtain

$$Z_1 = 4\pi V \beta^{-3} (2m)^{3/2} (h^{-3}) \int_0^\infty x^2 \exp(-x^2)\, dx. \qquad (4.6)$$

Here the integral of the function $x^2 \exp(-x^2)$ from 0 to ∞ is equal to $(\sqrt{\pi})/4$. Thus the final result is

$$Z_1 = V\left[\frac{2\pi mk_BT}{h^2}\right]^{3/2}, \qquad (4.7)$$

and the partition function is (following (3.33))

$$Z = \frac{\left\{V\left[\frac{2\pi mk_BT}{h^2}\right]^{3/2}\right\}^N}{N!}. \qquad (4.8)$$

We shall use the Stirling formula in the form $\text{Ln}(N!) = N \, \text{Ln} \, N - N$, for which $N! = (N/e)^N$, where e is the basis of the natural logarithm. Inserting this expression of $N!$ (correct for large N) one has the final expression of the partition function:

$$Z = \left(\frac{Ve}{N}\right)^N \left[\frac{2\pi m k_B T}{h^2}\right]^{3N/2}. \tag{4.9}$$

For the free energy $F = -k_B T \, \text{Ln} \, Z$, we get

$$F = -N k_B T \, \text{Ln} \left\{ \frac{eV}{N} \left[\frac{2\pi m k_B T}{h^2}\right]^{3/2} \right\}. \tag{4.10}$$

4.3.2 The internal energy, entropy and equation of state

Now, we can calculate three important quantities: energy, entropy and equation of state.

For the energy E, we use the thermodynamic relation $E = F - T(\partial F/\partial T)$ and state the well known expression of the ideal gas:

$$E = \frac{3}{2} N k_B T. \tag{4.11}$$

To derive this result, we develop the expression (4.10) of the free energy and write it as

$$F = -N k_B T \left[A + \frac{3}{2} \, \text{Ln} \, T) \right], \tag{4.12}$$

where $A = \text{Ln}(\frac{eV}{N}) + \frac{3}{2}\text{Ln}[\frac{2\pi m \, k_B}{h^2}]$. From $E = F - T(\frac{\partial F}{\partial T})$, one gets (4.11)

The internal energy is independent of the volume and is linearly related to the temperature.

The entropy is $S = -(\frac{\partial F}{\partial T})_{V,N}$, or

$$S = N k_B \left\{ (3/2) \, \text{Ln} \, T + \text{Ln}\left(\frac{eV}{N}\right) + 5/2 + (3/2) \, \text{Ln} \left[\frac{2\pi m k_B}{h^3}\right] \right\}. \tag{4.13}$$

This expression is compatible with that obtained in thermodynamics since it differs only by the constant $5/2 + (3/2) \, \text{Ln}[(2\pi m k_B)/h^3]$. In thermodynamics, one considers only differences in entropy, so that this constant does play any role. Note that this expression for S does not give $S = 0$ for $T = 0$. This is a consequence of the approximations of the classical case we made to derive the partition function (without any distinction between bosons and fermions).

The equation of state is $P = -(\frac{\partial F}{\partial V})_{T,N}$, or

$$P = \frac{Nk_BT}{V}. \tag{4.14}$$

This expression made possible the calculation of the value of the Boltzman constant k_B. For a mole (i.e. when the number of particles is equal to the Avogadro's number $N_A = 6.02 \times 10^{23}$) one has $P = RT/V$ (R is the ideal gas constant equal to $8.32\,\mathrm{J/K}$). One deduces $k_B = R/N_A = 1.38 \times 10^{-23}\,\mathrm{J/K}$.

Finally we calculate the chemical potential $\mu = \frac{\partial F}{\partial N}$. It is

$$\mu = -k_BT\left\{ \mathrm{Ln}\left(\frac{V}{N}\right) + \mathrm{Ln}\left[\frac{2\pi mk_BT}{h^2}\right]^{3/2} \right\}. \tag{4.15}$$

As mentioned above, in the classical limit, the chemical potential is negative (since $(V/N)[(2\pi mk_BT)/h^2]^{3/2} > 1$, as shown below).

4.3.3 *The classical limit*

In the preceding chapter, we have seen that the classical limit can be defined by the inequality $Z_1 \gg N$. Now, we are able to write explicitly this condition using the above expression of Z_1, (4.7),

$$V\left[\frac{2\pi mk_BT}{h^2}\right]^{3/2} \gg N. \tag{4.16}$$

We can express this inequality in two equivalent forms. If we consider the system at constant density N/V, we express a condition concerning the temperature,

$$k_BT \gg \frac{h^2}{2\pi m}\left(\frac{N}{V}\right)^{2/3}. \tag{4.17}$$

If we consider the system in constant temperature, the condition concerns the density

$$\frac{N}{V} \ll \left[\frac{2\pi mk_BT}{h^2}\right]^{3/2}. \tag{4.18}$$

In others terms, the classical limit is reached at low density, at high temperature, or at both.

Chapter 5

Some Problems

In order to have a better understanding of the concepts introduced in the preceding chapters we shall solve some problems. They are an integral part of the book and not merely exercises. It is strongly recommended that the reader does not skip this chapter.

5.1 The Quantum Harmonic Oscillator

We consider N particles with mass m, performing a harmonic motion in thermal contact with a reservoir at temperature T. Our aim is to calculate the thermal properties of these oscillators, namely the internal energy, the specific heat and the entropy. For this purpose, we shall determine the partition function Z. We suppose that an oscillator can move in the three dimensions of the space, and that the motion in each direction (x, y or z) is independent of the motion in the two others. This means that this ensemble of N oscillators in three dimensions is equivalent to $3N$ linear oscillators. We add a new assumption, namely that the positions of the oscillators are fixed. In this manner one can see them as distinguishable particles. Each oscillator is independent of the others, such that one can calculate Z from the partition function of a linear oscillator Z_1 by the relation (2.20), $Z = (Z_1)^{3N}$.

The different states of the linear oscillator are characterized by their energy e, which is

$$e = \left(n + \frac{1}{2}\right)\hbar\omega. \tag{5.1}$$

In (5.1), ω is the classical frequency of the oscillator multiplied by 2π. We recall that the potential energy of the oscillator is $K\omega^2 x^2/2$ (x is its displacement and K is a constant). The quantity n is an integer from $n = 0$ to infinity. There is one energy of each state such that in the sum of the partition function there is no distinction between sum on the microstates and sum on the energies.

The one-particle partition function is

$$Z_1 = \sum_n \exp\left[-\beta\left(n + \frac{1}{2}\right)\hbar\omega\right], \tag{5.2}$$

or

$$Z_1 = \exp\left(-\frac{\beta\hbar\omega}{2}\right)\sum_n \exp(-\beta n\hbar\omega). \tag{5.3}$$

Writing $x = \exp[-\beta\hbar\omega]$, the sum can be written as $[1 + x + x^2 + \cdots + x^n + \cdots]$ which converges to $1/(1-x)$ since $x < 1$. This gives

$$Z_1 = \frac{\exp\left(-\frac{\beta\hbar\omega}{2}\right)}{1 - \exp(-\beta\hbar\omega)}, \tag{5.4}$$

and for the total free energy of the N oscillators $F = -k_B T \operatorname{Ln} Z$ (with $Z = (Z_1)^{3N}$),

$$F = -3N\,k_B T \operatorname{Ln} Z_1 = 3N\left\{\frac{\hbar\omega}{2} + k_B T \operatorname{Ln}\left[1 - \exp(-\beta\hbar\omega)\right]\right\}. \tag{5.5}$$

To calculate the energy E, we use the relation $E = -\partial \operatorname{Ln} Z/\partial\beta = -3N(\partial \operatorname{Ln} Z_1/\partial\beta)$; it gives

$$E = 3N\left\{\frac{\hbar\omega}{2} + \hbar\omega\left[\exp\left(\frac{\hbar\omega}{k_B T}\right) - 1\right]^{-1}\right\}. \tag{5.6}$$

The specific heat[1] C is $C = dE/dT$, and

$$C = 3Nk_B\left(\frac{\hbar\omega}{k_BT}\right)^2 \exp\left(\frac{\hbar\omega}{k_BT}\right)\left[\exp\left(\frac{\hbar\omega}{k_BT}\right) - 1\right]^{-2}. \qquad (5.7)$$

It is possible to calculate the entropy in two different ways. The first by the expression (2.17) $S = -k_B\sum_s p_s \operatorname{Ln} p_s$, and the second by the thermodynamic relation $S = -\partial F/\partial T$. In the first way we shall calculate the entropy S_1 of an oscillator, and since the entropy is an additive quantity, we shall multiply the result by $3N$ for the entropy of all the oscillators. The probability p_s is given by (2.10), as

$$p_s = \frac{\exp\left(-\frac{E_s}{k_BT}\right)}{\sum_s \exp\left(-\frac{E_s}{k_BT}\right)} \quad \text{or} \quad p_s = \frac{\exp\left(-\frac{E_s}{k_BT}\right)}{Z_1}.$$

Writing the energy[2] of one oscillator as $E_s = (s + \frac{1}{2})\hbar\omega$, S_1 is given by

$$S_1 = -k_B\sum_s p_s \operatorname{Ln} p_s$$

$$= k_B\sum_n \frac{\exp\left[-\beta\left(s + \frac{1}{2}\right)\hbar\omega\right]}{Z_1}\left[\beta\left(s + \frac{1}{2}\right)\hbar\omega + \operatorname{Ln} Z_1\right] \qquad (5.8)$$

or by

$$S_1 = k_B\frac{\beta}{Z_1}\sum_n\left(s + \frac{1}{2}\right)\hbar\omega\exp\left[-\beta\left(s + \frac{1}{2}\right)\hbar\omega\right]$$

$$+ \frac{1}{Z_1}\operatorname{Ln} Z_1\sum_n\left\{\exp\left[-\beta\left(s + \frac{1}{2}\right)\hbar\omega\right]\right\}. \qquad (5.9)$$

The first term in the sum is minus the derivative of $\sum_n\{\exp[-\beta(s + \frac{1}{2})\hbar\omega] = Z_1$ relative to β and the second term is

---

[1]There is no distinction between the specific heat at constant volume and the specific heat at constant pressure, since we supposed that the frequency of an oscillator is independent of the volume and the pressure.

[2]We hope that the change in noting the quantum number n by the letter s will not trouble the reader.

merely $\text{Ln } Z_1$. One gets for S_1,

$$S_1 = k_B\left[-\frac{\beta\frac{\partial Z_1}{\partial \beta}}{Z_1} + \text{Ln } Z_1\right]. \tag{5.10}$$

Since $(\partial Z_1/\partial\beta)/Z_1 = \partial\text{Ln } Z_1/\partial\beta$, we have the final expression

$$S_1 = k_B\left[-\beta\frac{\partial \text{Ln } Z_1}{\partial \beta} + \text{Ln } Z_1\right]. \tag{5.11}$$

Now, from $\text{Ln } Z_1 = \beta\hbar\omega/2 + \text{Ln}[1 - \exp(-\beta\hbar\omega)]$, it is easy to calculate:

$$S = 3N\left\{\frac{\hbar\omega}{T}\left[\exp\left(\frac{\hbar\omega}{k_BT}\right) - 1\right]^{-1} - k_B\text{Ln}[1 - \exp(-\beta\hbar\omega)]\right\}. \tag{5.12}$$

The same result is obtained from the thermodynamic relation $S = -(\partial F/\partial T)_{V,N}$.

It is interesting to search for the limits of the expressions for E, C and S in the two cases $k_BT \ll \hbar\omega$ (low temperature limit) and $k_bT \gg \hbar\omega$ (high temperature limit).

5.1.1 *Low temperature limit*

If $k_BT \ll \hbar\omega$, one has either $\exp\left(\frac{\hbar\omega}{k_BT}\right) \gg 1$ or $\exp\left(\frac{-\hbar\omega}{k_BT}\right) \ll 1$. In this case, the expression for E becomes

$$E(k_BT \ll \hbar\omega) = 3N\hbar\omega\left[\frac{1}{2} + \exp\left(-\frac{\hbar\omega}{k_BT}\right)\right], \tag{5.13}$$

which goes to $3N\hbar\omega/2$ when T goes to zero.

The specific heat is given by (5.7) and its limit as $T \to 0$ is

$$C(k_BT \ll \hbar\omega) = 3Nk_B\left(\frac{\hbar\omega}{k_BT}\right)^2 \exp\left(-\frac{\hbar\omega}{k_BT}\right), \tag{5.14}$$

which goes to zero with T. To see this, it is necessary to consider the limit of $x^2\,e^{-x}$ as $x = \frac{\hbar\omega}{k_BT}$ goes to infinity. This expression is the product of two quantities, one going to zero e^{-x} and the other to infinity x^2 but the exponential approaches zero faster, such that the product goes also to zero. It is also possible to take the derivative of (5.13) relatively to T in order to get for (5.14).

For the limit of the entropy, we use the approximation $\text{Ln}(1-x) \approx -x$ for $|x| \ll 1$, and get

$$S(k_B T \ll \hbar\omega) = 3N\left[\frac{\hbar\omega}{T}\exp\left(-\frac{\hbar\omega}{k_B T}\right) + k_B \exp\left(-\frac{\hbar\omega}{k_B T}\right)\right],$$

(5.15)

which goes to zero with T, as expected for this entropy. Recall that $\exp(-1/x)/x$ goes to zero with x.

5.1.2 *High temperature limit*

In this case, $\frac{\hbar\omega}{k_B T} \ll 1$, one uses the approximation,

$$\exp\left(\pm\frac{\hbar\omega}{k_B T}\right) \approx 1 \pm \frac{\hbar\omega}{k_B T}$$

and has, for the energy

$$E(k_b T \gg \hbar\omega) = 3N k_B T,$$

(5.16)

and the specific heat

$$C(k_b T \gg \hbar\omega) = 3N k_B.$$

(5.17)

The entropy is the sum of two terms (see (5.12)). Using the approximation $\exp\left(\frac{\hbar\omega}{k_B T}\right) \approx 1 + \frac{\hbar\omega}{k_B T}$, the second term $-k_B \text{Ln}[1 - \exp(-\beta\hbar\omega)]$ can be written as $-k_B \text{Ln}(\beta\hbar\omega)$ or as $\text{Ln}\left(\frac{k_B T}{\hbar\omega}\right)$. Thus

$$S(k_b T \gg \hbar\omega) = 3N k_B\left[1 + \text{Ln}\left(\frac{k_B T}{\hbar\omega}\right)\right].$$

(5.18)

One sees that E and S increase with T, when C goes to a constant.

These results were obtained first by Einstein in 1907 and contain a historical importance since he was able to furnish an explanation of the decrease of the specific heat of solids with decreasing T. This experimental result was not explained by the classical statistical mechanics, which predict only the behavior given by the high temperature limit (that we can call the classical limit because the quantum behavior disappears; see Sec. 2.14). The thermal properties of solids were explained by postulating that atoms in solids behave like harmonic oscillators with a well defined frequency. Einstein was the first to show that it is necessary to use quantum mechanics. However, his success was only qualitative since the experimental decrease of C is not seen in the expression (5.14), but C is proportional to T^3. We shall see later in the second part of the book the solution to this problem.

5.2 The Polyatomic Ideal Gas

In the previous chapter, we calculated the properties of the monatomic ideal gas, i.e. a gas compound of single atoms. In this case the energy of an atom is its kinetic energy of translation. Now in the case of molecules with several atoms, there are, besides the kinetic energy of translation, the energies: a) due to vibrations of the atoms in the molecule; b) kinetic due to rotation. Thus the energy of one molecule is given by

$$E = E_t + E_{\text{vib}} + E_{\text{rot}}, \tag{5.20}$$

where $E_t = p^2/(2m)$ and the one-molecule partition function by

$$Z_1 = \sum \exp[-\beta(E_t + E_{\text{vib}} + E_{\text{rot}})], \tag{5.21}$$

$$Z_1 = \sum \exp(-\beta E_t)\exp(-\beta E_{\text{vib}})\exp(-\beta E_{\text{rot}}). \tag{5.22}$$

In (5.21) and (5.22), the sum is over all the microstates, when a microstate is defined by the five numbers: the three numbers n_x, n_y and n_z of the linear momentum, the quantum number n of the vibration energy and the quantum number corresponding to the z component of the angular momentum.

We suppose that different energies are not related, and write Z_1 as

$$Z_1 = \sum_t \exp(-\beta E_t) \sum_{\text{vib}} \exp(-\beta E_{\text{vib}}) \sum_{\text{rot}} \exp(-\beta E_{\text{rot}}), \tag{5.23}$$

or

$$Z_1 = Z_t Z_{\text{vib}} Z_{\text{rot}}. \tag{5.24}$$

As an example, we shall take the case of a molecule made of two identical atoms. The energy of vibration is given by the expression (5.1), and the energy of rotation is given by

$$E_{\text{rot}} = \frac{\hbar^2}{2I} K(K+1), \tag{5.25}$$

where I is the angular momentum of the molecule and K is a quantum number that is null or equal to a positive integer. Further there are $2K + 1$ states with the same energy corresponding to the z components of the angular momentum. These are the results of the quantum mechanics solution of the problem of the rotator. Z_t and Z_{vib} have been already calculated. Z_t is the one-particle partition function calculated in the preceding chapter (by expression (4.9)) with the difference that the mass m is now the sum of the two masses of the two atoms. Z_{vib} is given by the expression (5.4). It remains to calculate Z_{rot}.

$$Z_{\text{rot}} = \sum_K (2K+1) \exp\left[-\frac{K(K+1)\hbar^2}{2Ik_BT}\right]. \tag{5.26}$$

It is not possible to calculate Z_{rot} analytically and we look for approximations in the two limits $\hbar^2/(2Ik_BT) \ll 1$ and $\hbar^2/(2Ik_BT) \gg 1$. In the first case, the temperature is high enough such the important terms in the series are those with large values of K. One can replace the sum by an integral (with $K \gg 1$):

$$Z_{\text{rot}}\left(\frac{2Ik_BT}{\hbar^2} \gg 1\right) = \int_0^\infty 2K \exp\left[-\frac{K^2\hbar^2}{2Ik_BT}\right] dK. \tag{5.27}$$

The integral is calculated for K varying from 0 to infinite. To calculate the integral, we initiate a change in the variable, $x^2 = K^2\hbar^2/(2Ik_BT)$, and the integral becomes

$$Z_{\text{rot}}\left(\frac{2Ik_BT}{\hbar^2} \gg 1\right) = \frac{4Ik_BT}{\hbar^2} \int_0^\infty x \exp(-x^2)\, dx. \tag{5.37}$$

The integrand is the derivative of the function $\exp(-x^2)$ multiplied by $-1/2$ and the integral between the limits $[0, \infty]$ is equal to $\frac{1}{2}$. The final result is

$$Z_{\text{rot}}\left(\frac{2Ik_BT}{\hbar^2} \gg 1\right) = \frac{2Ik_BT}{\hbar^2}. \tag{5.38}$$

Now in the case where $2Ik_BT/\hbar^2 \ll 1$, the exponentials are small in the sum of (5.26) and one can take only the first two terms ($K = 0$

and $K = 1$):

$$Z_{\text{rot}}\left(\frac{2Ik_BT}{\hbar^2} \ll 1\right) = 1 + 3\exp\left(-\frac{\hbar^2}{2Ik_BT}\right). \qquad (5.39)$$

We have all the ingredients to calculate the thermal properties of this diatomic ideal gas. We see that we can have different regimes following the various conditions: for the rotation when $K^2\hbar^2/(2I\,k_BT) \ll 1$ or $K^2\hbar^2/(2I\,k_BT) \gg 1$, for the vibration when $k_BT \ll \hbar\omega$ or $k_BT \gg \hbar\omega$, with the condition (4.15), $k_BT \gg (h^3/2\pi m)(N/V)^{2/3}$, of the ideal gas. We shall consider the case of the high temperature limit in admitting that when (4.15) is satisfied, then $K^2\hbar^2/(2I\,k_BT) \ll 1$ and $k_BT \gg \hbar\omega$. The energy and the specific heat at constant volume can be calculated using the preceding results (expressions (4.11) for the energy of translation, (5.16) for the energy of vibration and (5.38) for the partition function of the rotation). This gives

$$E = \frac{3}{2}N\,k_BT + 3N\,k_BT + N\,k_BT = \frac{11}{2}N\,k_BT, \qquad (5.40)$$

$$C_V = \frac{11}{2}Nk_B. \qquad (5.41)$$

In general, for an ideal gas at room temperature when (4.15) is satisfied, one has $k_BT \ll \hbar\omega$ such that the contribution of the vibration to the specific heat is negligible and $C_V = (5/2)Nk_B$, as indicated in the textbook of thermodynamics for a diatomic gas.

5.3 Bosons and Fermions in a Two-Level System

One considers N particles with $2M$ possible states, among them there are M states with energy equal to zero and M states with energy e. There are two possible cases: the particles are bosons or they are fermions. In the second case, one must have $2M \geq N$ since a state cannot be occupied by more than one fermion. We wish to know the thermal properties of this system of N particles. We shall see that the properties of such a system can vary much depending of the nature of the particles whether they are bosons or fermions.

To solve this problem, we do not use the partition function, but with the chemical potential. We begin by calculating the chemical potential using the expression (3.22b) that gives the number of particles

$$N = \sum_i \{\exp[\beta(e_i - \mu)] \pm 1\}^{-1}. \qquad (3.22b)$$

Again, we recall that the sum is over the states. In the present cases there are only two possible energies, but $2M$ states and N is given by

$$N = n_1 + n_2,$$

$$n_1 = M[\exp(-\beta\mu) \pm 1]^{-1},$$

$$n_2 = M\{\exp[\beta(e - \mu)] \pm 1\}^{-1}.$$

Writing $x = \exp(-\beta\mu)$ and $A = \exp(\beta e)$, one has an equation to extract the chemical potential as a function of T and N,

$$M[(x \pm 1)^{-1} + (Ax \pm 1)^{-1}] = N. \qquad (5.42)$$

5.3.1 *The particles are bosons*

For this case, (5.42) becomes

$$M[(x - 1)^{-1} + (Ax - 1)^{-1}] = N, \qquad (5.43)$$

and can be written as follows:

$$\frac{M[(Ax - 1) + (x - 1)]}{(x - 1)(Ax - 1)} = N, \qquad (5.43b)$$

or

$$M[Ax + x - 2] = N(x - 1)(Ax - 1), \qquad (5.43c)$$

or

$$\frac{N}{M}Ax^2 - (A + 1)\left(\frac{N}{M} + 1\right)x + \left(2 + \frac{N}{M}\right) = 0. \qquad (5.44)$$

Before solving this equation, we shall look to the solutions for $T \to \infty$ and $T \to 0$. In the first case, $A = \exp(e/(k_B T)) \to 1$ and (5.44) becomes

$$\frac{N}{M}x^2 - 2\left(\frac{N}{M} + 1\right)x + \left(2 + \frac{N}{M}\right) = 0. \qquad (5.45)$$

This equation has two solutions; they are $x_1 = (2 + N/M)(M/N)$ and $x_2 = 1$. If $x = 1$, $n_1 = M/(x - 1)$ goes to infinity and the solution is not acceptable. Inserting the correct value for x_1 in the above expressions of n_1 and n_2 gives $n_1 = n_2 = N/2$ and $\mu = -k_B T \operatorname{Ln}(1 + 2M/N)$.

In the second case $(T \to 0) A \to \infty$, neglecting the third term in (5.45) and dividing by A, Eq. (5.44) becomes

$$\left(\frac{N}{M}\right) x^2 - \left(\frac{N}{M} + 1\right) x = 0. \tag{5.46}$$

The solution is $x = 1 + M/N$ since $x > 0$. This gives $n_1 = N$, $n_2 = 0$ (since A is infinite) and $\mu = -k_B T \operatorname{Ln}(1 + M/N)$, which is always negative as it stands for bosons. This analysis shows that regardless of the number of states and the number of particles, at low temperatures all the particles are in state 1 and the energy is null; while at high temperatures, the particles are equally distributed in the two levels and the energy is $E = Ne/2$.

The solution of the general problem (whatever values N and M take) is cumbersome and we direct our discussion to solutions for the two following cases: (i) $M = N$; (ii) $N \ll M$.

Case 1: N=M.
Equation (5.44) becomes

$$Ax^2 - 2(A + 1)x + 3 = 0, \tag{5.47}$$

with two solutions $x = (1 + 1/A) \pm (1/A)\sqrt{A^2 - A + 1}$. We can verify that the solutions which give the known results in the limits $T \to 0$ and $T \to \infty$, have the sign $+$. We calculated numerically n_2, the number of particles in level 2, and plotted in Fig. 5.1 as a function of the temperature in the, manner of Chapter 1, $e/k_B = 10°K$. The energy E as a function of T has the same variation as n_2, since $E = en_2$.

Case 2: N≪M.
Equation (5.44) can be written as (neglecting N/M relative to 1 and 2)

$$\frac{N}{M} Ax^2 - (A + 1)x + 2 = 0. \tag{5.48}$$

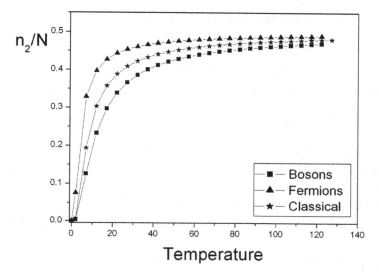

Figure 5.1. Variation of the number of particles in the state with energy e, in the three cases: bosons, fermions and classical particles.

The positive solution[3] is

$$x = \frac{M}{2NA} \left\{ (A+1) + \left[(A+1)^2 - \frac{8NA}{M} \right]^{0.5} \right\}.$$

In the square root, one has[4] $(A+1)^2 \gg 8NA/M$ and the solution is with a very good approximation $x = (M/N)(1 + 1/A)$. One remarks that in this case, x and xA are much larger than one and the expressions for n_1 and n_2 are merely $n_1 = M/x = N/(1 + 1/A)$ and $n_2 = M/(xA)$. The energy is

$$E = Ne[1 + \exp(\beta e)]^{-1}, \qquad (5.49)$$

and the chemical potential is calculated from the expression of $x = \exp(-\mu/(k_B T)) = (M/N)(1 + 1/A)$,

$$\mu = -k_B T \operatorname{Ln} \left[\frac{M}{N} (1 + \exp(-\beta e)) \right]. \qquad (5.50)$$

Note that μ is negative and goes to zero with T.

---

[3]We opt again for the positive solution, using the same reason as above.
[4]We recall that A varies from 1 to infinity, and consequently this inequality is always fulfilled when $M \gg N$.

5.3.2   *The particles are fermions*

The equation bringing forth x is now (from (5.42))

$$M[(x+1)^{-1} + (Ax+1)^{-1}] = N, \qquad (5.51)$$

or (following the same steps as in the case for the bosons)

$$\frac{N}{M}Ax^2 + (A+1)\left(\frac{N}{M}-1\right)x - \left(2-\frac{N}{M}\right) = 0. \qquad (5.52)$$

By the same method as above for the bosons, it is possible to show that for $T \to 0 (A \to \infty)$, one has $n_1 \to N$, $n_2 \to 0$, and $x = \frac{M}{N} - 1$. In the limit of high temperatures $(T \to \infty, A \to 1)$, one has $n_1 = n_2 = N/2$ and $x = 2M/N - 1$.

As in the case of bosons, we shall make explicit calculations for simple cases:

Case 1: M = N.

When $M = N$, Eq. (5.52) simplifies, and becomes merely

$$Ax^2 - 1 = 0, \qquad (5.53)$$

or

$$x = A^{-1/2} \qquad (5.54)$$

and $\mu = e/2$. The chemical potential is constant and located at equal distances from the two levels.

We plotted in Fig. 5.1 the variations of n_2, with T with the same condition as above, i.e. $e/k_B = 10\,^\circ$K. One sees that, in the case of fermions, n_2 increases much faster than it is in the case of bosons, as a consequence of the Pauli principle.

Case 2: $N \ll M$.

Equation (5.53) becomes

$$\frac{N}{M}Ax^2 - (A+1)x - 2 = 0, \qquad (5.55)$$

and the positive solution is

$$x = \frac{M}{2NA}\left\{(A+1) + \left[(A+1)^2 + \frac{8NA}{M}\right]^{0.5}\right\}.$$

As done for Sec. 5.3.1, neglecting insignificant terms of $8NA/M$ relative to $(A+1)^2$ gives the same solution as it is for the bosons,

$x = (M/N)(1 + 1/A)$. A conclusion is reached, that when $N \ll M$, there is no distinction between bosons and fermions. We consider the particles as classical particles, as in the case of the ideal gas. In Fig. 5.1, the variation of n_2 is plotted for the classical situation. It is worth while to give a detailed analysis of the classical situation.

5.3.3 *Classical particles*

Our objective is to calculate the free energy of the system in the classical case $N \ll M$. The method is general and in the classical case, it is possible to make easily explicit calculations. We know the energy

$$E = \frac{Ne}{1 + \exp(\beta e)} = -\frac{\partial \operatorname{Ln} Z}{\partial \beta},$$

and the chemical potential

$$\mu = -k_B T \operatorname{Ln}\left[\frac{M}{N}(1 + \exp(-\beta e))\right] = \left(\frac{\partial F}{\partial N}\right)_T = \frac{\partial(-k_B T \operatorname{Ln} Z)}{\partial N}.$$

In other words, we have the two derivatives of $\operatorname{Ln} Z$, from which it is possible to find the partition function Z, as a function of T and N. From the first, $\partial \operatorname{Ln} Z/\partial \beta = -Ne/[1 + \exp(\beta e)]$, one can write

$$\operatorname{Ln} Z = -\int d\beta \frac{Ne}{1 + \exp(\beta e)} + K(N), \qquad (5.56)$$

where $K(N)$ is an unknown function of the number of particles N. To calculate the integral, we make the change of variables $x = \beta e$ and $u = 1 + \exp(x)$. We then have $du = \exp(x)\, dx$ or $dx = du/(u-1)$, taking into account that $\exp(x) = u - 1$. $\operatorname{Ln} Z$ is then

$$\operatorname{Ln} Z = -N \int \frac{du}{u(u - 1)} + K(N) \qquad (5.57)$$

Writing $1/[u(u - 1)] = -1/u + 1/(u - 1)$, one gets for the integral

$$\int \frac{du}{u(u - 1)} = -\int \frac{du}{u} + \int \frac{du}{u - 1} = \operatorname{Ln}\left[\frac{u - 1}{u}\right]. \qquad (5.58)$$

Finally

$$\operatorname{Ln} Z = N \operatorname{Ln}[1 + \exp(-\beta e)] + K(N). \qquad (5.59)$$

The second derivative is

$$\frac{\partial(-k_BT\operatorname{Ln}Z)}{\partial N} = -k_BT\operatorname{Ln}\left[\frac{M}{N}(1+\exp(-\beta e))\right]. \tag{5.60}$$

It can be written as

$$\left(\frac{\partial\operatorname{Ln}Z}{\partial N}\right)_T = \operatorname{Ln}M - \operatorname{Ln}N + \operatorname{Ln}[1+\exp(-\beta e)], \tag{5.61}$$

giving

$$\operatorname{Ln}Z = -\int \operatorname{Ln}N\,dN + \operatorname{Ln}M\int dN$$
$$+ \operatorname{Ln}[1+\exp(-\beta e)]\int dN + P(T), \tag{5.62}$$

where $P(T)$ is an unknown function of T.

Recalling, that the integral of $\operatorname{Ln}N$ is $(N\operatorname{Ln}N - N)$, (5.62) can be written as

$$\operatorname{Ln}Z = N\operatorname{Ln}[1+\exp(-\beta e)] + N(\operatorname{Ln}M - \operatorname{Ln}N + 1) + P(T). \tag{5.63}$$

Comparing (5.59) and (5.63), conclude that $P(T) \equiv 0$ and that $K(N) = N[\operatorname{Ln}(M/N) - 1]$. The result for F is

$$F = -k_BTN\left\{\operatorname{Ln}[1+\exp(-\beta e)] + \operatorname{Ln}\left(\frac{M}{N}\right) + 1\right\}. \tag{5.64}$$

From (5.64) it is possible to calculate the entropy and its limits for $T \to 0$ and $T \to \infty$.

Finally, one can formulate the partition function Z with the help of the one-particle partition function Z_1. By definition, Z_1 is given by

$$Z_1 = M[1+\exp(-\beta e)]. \tag{5.65}$$

From (5.64), the logarithm of Z is

$$\operatorname{Ln}Z = \frac{F}{-kT} = N\left\{\operatorname{Ln}[1+\exp(-\beta e)] + \operatorname{Ln}\left(\frac{M}{N}\right) + 1\right\}, \tag{5.66}$$

or

$$\operatorname{Ln}Z = N\operatorname{Ln}[1+\exp(-\beta e)] + N\operatorname{Ln}M - N\operatorname{Ln}N + N, \tag{5.67}$$

or

$$\operatorname{Ln}Z = N\{\operatorname{Ln}[1+\exp(-\beta e)] + \operatorname{Ln}M\} - N\operatorname{Ln}N + N, \tag{5.68}$$

or

$$\operatorname{Ln}Z = \operatorname{Ln}\{M^N[1+\exp(-\beta e)]^N\} - N\operatorname{Ln}N + N \tag{5.69}$$

Now we shall use Stirling's formula, $\text{Ln } N! \approx N \text{ Ln } N - N$, and finally

$$\text{Ln } Z = \text{Ln}\left\{ \frac{M^N[1 + \exp(-\beta e)]^N}{N!} \right\}, \tag{5.70}$$

$$Z = \frac{M^N[1 + \exp(-\beta e)]^N}{N!} = \frac{(Z_1)^N}{N!}. \tag{5.71}$$

One can compare the results of this section with those of the example that we took in Chapters 1 and 2 (N particles in a two-level system). In particular, the presence of $N!$ in the denominator of Z is due to the fact that in the present case, it is not possible to distinguish between the particles.

Important remark. From the preceding chapters, we see that there are four types of particles:

The indistinguishable quantum particles with $z_i = \exp[-\beta(e_i - \mu)]$ are

(1) Fermions: The grand partition function is given by $Z_G = \Pi_i (1 + z_i)^{+1}$;
(2) Bosons: The grand partition function is given by $Z_G = \Pi_i (1 - z_i)^{-1}$.

Those with N classical particles, (where Z_1 is the one particle partition function) are

(3) Distinguishable: The partition function is $Z = (Z_1)^N$.
(4) Indistinguishable: The partition function is $Z = (Z_1)^N/N!$.

Particles behave classically only in particular circumstances (for example, the ideal gas and the preceding example).

5.4 The Magnetic Chain

We consider a chain made of N magnetic dipoles. A dipole is attached to another dipole to form a magnetic chain. The chain is located in a plan and each dipole b can be directed along four directions $+x$, $-x$, $+y$ and $-y$. One end of the chain is fixed at the point $x = 0$, $y = 0$, when the other end is free and a magnetic field H is applied in the $+x$ direction. The energy of a dipole is equal to $-bH \cos \delta$ when

δ is the angle between the vector magnetic moment and the vector magnetic field. The energy of the dipole is zero if it is along the $\pm y$ directions, $e_1 = -bH$ if it points to the $+x$ direction and $e_2 = bH$ for the $-x$ direction. Furthermore, to make the problem simple, we suppose that the state of one dipole is not influenced by the state of its two neighbors, and so two dipoles can be superposed. By this simplification, the dipoles do not interact. Since the position of one dipole in the chain is well defined, the dipoles are distinguishable units. The partition function is $Z = Z_1^N$. One dipole has four possible states corresponding to the four directions that it can take. The energy of these four states are $E_1 = bH$, $E_2 = -bH$, $E_3 = 0$ and $E_4 = 0$. Thus,

$$Z_1 = \exp(-\beta bH) + \exp(\beta bH) + 2 = 2[1 + \text{ch}(\beta bH)]. \qquad (5.66)$$

The energy is calculated with the help of the relation $E = -\partial \operatorname{Ln} Z / \partial \beta$. Hence,

$$E = -\frac{NbH\operatorname{sh}(\beta bH)}{1 + \operatorname{ch}(\beta bH)}. \qquad (5.67)$$

At high temperature $(\beta \rightarrow 0)$, $\text{ch}(\beta bH) \rightarrow 1$ and $\text{sh}(\beta bH)$ goes to zero, i.e. E goes also to zero. Now at low temperature, $(\beta \rightarrow \infty)$ and both $\text{sh}(\beta bH)$ and $\text{ch}(\beta bR)$ are very large and practically equal to $(1/2)\exp(\beta bH)$. Thus, E goes to a constant value, $-NbH$.

The entropy is derived from the expression $S = (E - F)/T = E/T + k_B \operatorname{Ln} Z$, that it is

$$S = Nk_B \operatorname{Ln}\{2[1 + \text{ch}(\beta bH)]\} - \frac{NbH}{T}\frac{\text{sh}(\beta bH)}{1 + \text{ch}(\beta bH)}, \qquad (5.68)$$

$$S = Nk_B\{\operatorname{Ln} 2 + \operatorname{Ln}[1 + \text{ch}(\beta bH)]\} - \frac{Nk_B(\beta bH)\text{sh}(\beta bH)}{1 + \text{ch}(\beta bH)}.$$

$$(5.69)$$

Now we calculate the limits of S at high and low temperatures. For high $T(\beta \rightarrow 0)$, one gets $S = Nk_B \operatorname{Ln} 4$. This result is general for a system of N particles with finite states. At low temperature, we know that S goes to zero. We shall verify this basic property of the entropy. We write S as $(x = \beta bH)$

$$S = Nk_B\{\text{Ln}\,2 + \text{Ln}[1 + \text{ch}(x)]\} - Nk_B x \frac{\text{sh}(x)}{1 + \text{ch}(x)},$$

and look for the limit of S when $x \to \infty$. Putting $\text{ch}(x) \approx \text{sh}(x) \approx (1/2)\exp(x)$ for $x \to \infty$, one has

$$S \to Nk_B\left\{\text{Ln}\,2 + \text{Ln}\left[\frac{1}{2}\exp(x)\right] - x\right\}$$
$$= Nk_B[\text{Ln}\,2 - \text{Ln}\,2 + x - x] = 0. \qquad (5.70)$$

The magnetization of the chain can be calculated using the magnetic free energy F_M, via $M = -\partial F_M/\partial H$. The free energy is $-k_B T N \,\text{Ln}\, Z_1$ and one gets

$$M = Nb\frac{\text{sh}(\beta bH)}{1 + \text{ch}(\beta bH)}. \qquad (5.71)$$

It is also possible to calculate M through the following expression

$$M = bN(x+) - bN(x-), \qquad (5.72)$$

when $N(x+)$ is the number of dipoles in the $x+$ direction and $N(x-)$ the number in the $x-$ direction. The probability $p(x+)$ for a dipole to be in the $x+$ direction is given by $\exp(\beta bH)/Z_1$, and the probability $p(x-)$ for a dipole to be in the $x-$ direction is $\exp(-\beta bH)/Z_1$. The number of dipoles in the $x+$ direction is $Np(x+)$ and the number of dipoles in the $x-$ direction is $Np(x-)$. One gets the same result, (5.71).

At low temperature $(\beta \to \infty)$ or large fields $(H \to \infty)$, M tends towards Nb since $\text{sh}(\beta bH)$ and $\text{ch}(\beta bH)$ both approach $(1/2)\exp(\beta bH) \gg 1$. The magnetization, fetches its maximum value. In this case, the chain is completely linear since all the dipoles are in the direction $x+$. However, at high temperature and low magnetic fields, M tends toward zero. This means that, in this limit, the chain has a complicated shape, with the same number of dipoles in all directions. To show, recall that for $T \to \infty$, $\text{ch}(\beta bH)$ goes to $\frac{1}{2}$ and $\exp(\pm\beta bH)$

goes to $1 \pm (\beta b H)$, one has

$$p(x+) \approx \frac{1 + \beta b H}{4},$$

$$p(x-) \approx \frac{1 - \beta b H}{4},$$

$$p(y+) = p(y-) = \frac{1}{4},$$

all of which approach $1/4$ at high temperature.

As the temperature changes from zero to infinity, the chain evolves from a linear chain to a complicated closed loop, and the two extremal points of the chain are near each another.

The solution we present portrays a simplified view of the chain. A more comprehensive solution is much more difficult since one has to take into account the fact that two dipoles cannot overlap.

PART II
Applications

In the first part, we developed the basic principles of statistical mechanics. Now we use them to study some real situations. They are physical problems which were studied experimentally and received complete explanation by the methods developed in the first part.

We begin with three cases of bosons: the photons, the phonons and the boson gas at low temperature. In the first two cases, the number of particles is not fixed and their mass is zero, and the third consists of a gas of massive particles. The phonon is a new particle that the majority of readers have not encountered. To understand its properties, one does need to know more than the equivalence between waves and particles that one learns in quantum mechanics. The fermions that we study are electrons which are very important in solid state physics. We deal with two cases: electrons in metals and electrons in semiconductors. In the frame of this book, the particles we consider are independent, without explicit interaction, and the problems that we solve are the simplest problems; in most of the cases in physics, it is not possible to ignore their interaction. For them, sophisticated methods were developed which are beyond the scope of this book.

Chapter 6

The Gas of Photons:
The Black Body Radiation

It is well known that when a body is heated it begins to emit light, and the spectrum of the light emitted, i.e. the distribution of wavelengths, is dependent on the temperature. If one uses the naked eye to evaluate the wavelength or the color, when the temperature of the body increases, one begins to see a reddish color, after that a bright red light appears, and further a white light. This radiation is due to the emission of electromagnetic waves or photons. The origin of the emission is the motion of the atoms of the body, which are moving because of heat. In the theory of electromagnetism, an electric charge moving in accelerated motion emits an electromagnetic radiation. In the solid, the charged particles (electrons and nuclei) move in a disordered motion such that they emit photons. In quantum mechanics, it is taught that phenomena associated with light can be described by two complementary ways — as waves or as particles. Depending of the situation, one is more appropriate than the other. In the present case, we shall use the particle aspect, without ignoring the other.

In this chapter we shall consider two phenomena which apparently are not directly related — the photon gas and light emission by solids. In the case of the photon gas, i.e. photons inside a container, one is interested in knowing, besides the regular thermodynamic properties of a gas (energy, entropy, equation of state *etc.*), what are the wave components of the light spectrum. On the other hand, the relevant point in the light emission of solids is precisely their light spectrum. One important case is the light emission by what it is called the black body. Such a body appears black because it absorbs all radiations, whatever their wavelengths are. There is a direct relation between the light spectrum of a photon gas and that of a black body.

To be able to study precisely a photon gas, one has to define the conditions in which it takes place. For this purpose, one makes inside a body a cavity with a simple shape. Because the body is heated, the inner walls of the cavity emit photons and absorb at the same time photons. The only condition that we put on the body is that it can withstand the temperature. At the equilibrium, one has a gas of photons in a well defined volume (that of the cavity) and a well defined temperature. The energy of the gas is also well defined — there is no energy loss nor energy gain — since the cavity is closed.

The properties of the photon gas are independent of the shape of the cavity and the material of the walls. For the time being we shall take these properties as well verified experimental facts. Below we shall give a physical argument to show that the properties of the photon gas are independent of the shape and the material.

One asks: what is the energy of the photon gas? What is the spectrum of the radiation trapped in the cavity?

But how it is possible to answer experimentally these questions, since the gas is trapped in a close cavity without contact with the external world? To be able to make measurements on this photon gas, one makes a small aperture in the body such that a small quantity of radiation can escape outside. This hole is so small that it does not perturb the photon gas.

The radiation spectrum of the photon gas is defined as $S(\lambda)$, giving the energy of radiation between λ and $\lambda + d\lambda$ ($d\lambda$ is a small interval of wavelength). $S(\lambda)$ goes to zero with the wavelength, reaches a maximum, and decreases toward zero at long wavelengths. Experimentally, one obtains that E is proportional to the product VT^4. Furthermore, it was found that the product $\lambda_m T$ is constant, when λ_m is the wavelength of maximum intensity in the spectrum.

Before calculating the energy and the spectrum, one recalls the properties of the photon. First, as a particle it is defined by its moment p, and its energy $E = pc$ (c is the light velocity). As a wave it has a wave vector k, a wavelength $\lambda = 2\pi/k$ and a frequency ν or $\omega = 2\pi\nu$. Between these quantities, one has the following relations:

$$k = \frac{2\pi}{\lambda} = \frac{2\pi\nu}{c} = \frac{\omega}{c}\lambda = \frac{c}{\nu}. \tag{6.1}$$

The relations between the two series of properties of a photon, as a particle and as a wave, are

$$E = pc = h\nu = \frac{hkc}{2\pi} = \hbar kc, \tag{6.2}$$

$$p = \hbar k. \tag{6.3}$$

In (6.2) and (6.3) h is the Planck constant and $\hbar = h/(2\pi)$.

6.1 The Energy and the Energy Spectrum

We recall the formula for the energy provided in Chapter 3

$$E = \sum_r n_r e_r, \tag{3.23b}$$

when n_r is the mean number of particles in the state r, and e_r is the energy of the state r. The number n_r is given by (3.24) fixing $\mu = 0$,

$$n_r = [\exp(\beta e_r) - 1]^{-1}. \tag{6.4}$$

The chemical potential is null since the number of photons is not fixed. There is permanent production and absorption of photons: their number is not fixed. The photons are bosons since their spin is 1 and thus there is no restriction on the number of photons in a given state.

The sum (3.23b) is over the states, but the sum over the energies is

$$E = \sum g(e_r) n_r e_r, \tag{6.5}$$

where $g(e_r)$ is the density of states, the number of states with energy e_r. However, in Chapter 4 on the density of states, we mentioned that the energy levels are so close to each other that one can replace the sum (6.4) by an integral

$$E = \int_0^\infty E\, g(E) n(E)\, dE, \tag{6.6}$$

with $n(E) = [\exp(\beta E) - 1]^{-1}$.

The density of states g(p), as a function of the momentum p, was calculated in the Chapter 4 (Eq. (4.4)), as

$$g(p)\, dp = s_z (4\pi V p^2\, dp)/h^3.$$

To transform $g(p)$ into $g(E)$, one has to recall that $E = pc$ or $p = E/c$. This gives,[1]

$$g(E)\, dE = \frac{8\pi V}{c^3 h^3} E^2\, dE. \tag{6.7}$$

We added a factor of 2 (s_z is equal to 2, because the two possible polarizations of the photon which is a transversal wave).

Eq. (6.6) is now

$$E = \frac{8\pi V}{c^3 h^3} \int_0^\infty E^3 [\exp(\beta E) - 1]^{-1}\, dE. \tag{6.8}$$

The limits of the integration are 0 and ∞. To calculate the integral, one multiplies and divides the integral by $\beta^{-4} = (k_B T)^4$. The integral becomes

$$I = \int_0^\infty E^3 [\exp(\beta E) - 1]^{-1}\, dE$$

$$= \int_0^\infty (k_B T)^4 \left(\frac{E}{k_B T}\right)^3 \left[\exp\left(\frac{E}{k_B T}\right) - 1\right]^{-1} d\left(\frac{E}{k_B T}\right),$$

or writing $x = \frac{E}{k_B T}$, shifting $(k_B T)^4$ outside, and taking the same integration limits $[0, \infty]$,

$$I = (k_B T)^4 \int_0^\infty x^3 [\exp(x) - 1]^{-1}\, dE = \frac{\pi^4}{15}(k_B T)^4,$$

where $\pi^4/15$ is the value of the definite integral. Now the energy is easily calculated from Eq. (6.8) by replacing the integral I with its value:

$$E = \frac{8\pi^5 k_B^4 V}{15 h^3 c^3} V T^4. \tag{6.9}$$

---

[1]The spin of the photon is 1. In principle s_z should be equal to 3. However, for a massless particle there are only two spin projection, corresponding to the two possible polarizations of the light.

It is common practice to write E as

$$E = \frac{4\sigma}{c} V T^4, \tag{6.10}$$

where $\sigma = 2\pi^5 k_B^4/(15h^3c^2)$ is the Stefan constant. It is equal to $5.67 \ 10^{-8}$ MKS. The experimental determination is in very good agreement with its theoretical formula.

The energy spectrum can be calculated from the expression (6.8) of the energy written as

$$E = \int S(E)\, dE,$$

where $S(E)\, dE$ is the quantity of energy between the two values E and $E + dE$. Because of the wave–particle duality, it is possible to translate the expression of $S(E)\, dE$ as a function of the wavelength, since one has $E = pc = \hbar kc = hc/\lambda$.

From (6.8)

$$S(E)\, dE = \frac{8\pi V}{c^3 h^3} E^3 [\exp(\beta E) - 1]^{-1}\, dE,$$

and designating $E = \frac{hc}{\lambda}$ and $dE = \frac{hc}{\lambda^2 d\lambda}$, one gets

$$S(\lambda)\, d\lambda = \frac{8\pi V h c}{\lambda^5} \left[\exp\left(\frac{hc}{\lambda k_B T}\right) - 1\right]^{-1} d\lambda. \tag{6.11a}$$

Here $S(\lambda)$ is the radiation spectrum of the photon gas. The quantity

$$K(\lambda) = \frac{S(\lambda)}{V} = \frac{8\pi hc}{\lambda^5} \left[\exp\left(\frac{hc}{\lambda k_B T}\right) - 1\right]^{-1} \tag{6.11b}$$

has the dimension of energy per volume per wavelength, and can be seen as the energy density of the photon gas.

The quantity $K(\lambda)$ is plotted for several temperatures in Fig. 6.1. These theoretical curves are in a perfect agreement with the experimental curves. The spectrum $S(\lambda)$ varies very strongly with the temperature that we have used a logarithmic scale to plot the quantity $K(\lambda)$.

To find the law of the constant product $\lambda_m T$, one derives the function $S(\lambda)$ relative to λ, and writes that the derivative is null

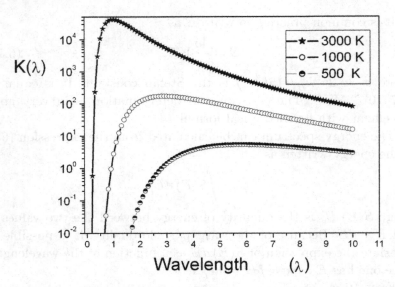

Figure 6.1.   Energy density spectrum of the photon gas, as a function of the wavelength (in microns), for several temperatures. The energy spectrum is strongly dependent on the temperature.

for λ_m. First, one writes $S(\lambda) = 8\pi hc/f(\lambda)$, and has $dS/d\lambda = -(8\pi hc)(df/d\lambda)/f^2$. Writing $dS/d\lambda = 0$ is equivalent as writing $df/d\lambda = 0$. The function $f(\lambda)$ is equal to $\lambda^5[\exp(hc/\lambda(kT)) - 1]$, and one has

$$\frac{df}{d\lambda} = 5\lambda^4\left[\exp\left(\frac{hc}{\lambda_m k_B T}\right) - 1\right] - \lambda_m^3\left(\frac{hc}{k_B T}\right)\exp\left(\frac{hc}{\lambda_m k_B T}\right) = 0,$$
(6.12a)

$$\frac{df}{d\lambda} = \lambda_m^4\left\{5\left[\exp\left(\frac{hc}{\lambda_m kT}\right) - 1\right] - \left(\frac{hc}{\lambda_m k_B T}\right)\exp\left(\frac{hc}{\lambda_m k_B T}\right)\right\} = 0.$$
(6.12b)

One sees that in (6.12b), one can solve the equation in the brackets by taking the quantity $hc/(\lambda_m k_B T)$ as the unknown. The solution will be ascertained for $hc/(\lambda_m k_B T)$, showing that the product $\lambda_m T$ is constant.

All the preceding results were obtaining by Planck in 1909 and they were the first steps toward quantum mechanics.

6.2 The Free Energy and the Entropy

The expression (3.13), applied to the bosons, gives the free energy of a gas of bosons with an undetermined number of particles.

$$F = -k_B T \sum_I -\text{Ln}[1 - \exp(-\beta e_i)]. \tag{3.13}$$

Passing to the energy level continuum, and taking into account the density of states, one can write

$$F = k_B T \int_0^\infty g(E) \, \text{Ln}[1 - \exp(-\beta E)] \, dE. \tag{6.13}$$

The expression for $g(E)$ is given by (6.7), thus

$$F = \frac{8\pi k_B T V}{h^3 c^3} \int_0^\infty E^2 \, \text{Ln}\left[1 - \exp\left(-\frac{E}{k_B T}\right)\right] dE. \tag{6.14a}$$

We repeat the trick used above to transform the integral, multiplying and dividing it by $(k_B T)^3$. This transforms the integral into a function of the group $E/(k_B T)$ alone and it becomes a definite integral.

$$F = \frac{8\pi k_B T V}{h^3 c^3} (k_B T)^3 \int_0^\infty \left(\frac{E}{k_B T}\right)^2 \text{Ln}\left[1 - \exp\left(-\frac{E}{k_B T}\right)\right] d\left(\frac{E}{k_B T}\right). \tag{6.14b}$$

This shows that F is proportional to T^4. The exact result is

$$F = -\frac{4\sigma}{3c} V T^4. \tag{6.15}$$

Formally, the negative sign comes from the fact that the logarithm in (6.14a) is negative, since $[1 - \exp(-E/(k_B T))]$ is smaller than 1.

From the free energy, one can calculate all the thermodynamic properties. The calculations are easy and we show only the results that the reader can check.

The entropy $S = -(\partial F/\partial T)_V = \frac{16\sigma}{3c} V T^3$, and the pressure $P = -(\partial F/\partial V)_T$ is equal to $\frac{4\sigma}{3c} T^4$. Also, $PV = E/3$.

The specific heat at constant volume is $C_V = (\partial E/\partial T)_V = (16\sigma/c) V T^3$.

The mean number of photons is given by

$$N_{\text{mean}} = \sum_r n_r = \sum_r [\exp(\beta e_r) - 1]^{-1}. \tag{6.16}$$

The sum is over the states and passing the sum over to the energy, one has

$$N_{\text{mean}} = \int_0^\infty g(E) \left[\exp\left(\frac{E}{k_B T}\right) - 1\right]^{-1} dE, \qquad (6.17)$$

or, using the above expression of $g(E)$,

$$N_{\text{mean}} = \frac{8\pi V}{h^3 c^3} \int_0^\infty E^2 \left[\exp\left(\frac{E}{k_B T}\right) - 1\right]^{-1} dE. \qquad (6.18)$$

Using the same trick as above, one multiplies the above expression by $(k_B T)^3$ and divides by the same quantity, thus one can write (6.18) as follows:

$$N_{\text{mean}} = \frac{8\pi V k_B^3 T^3}{h^3 c^3} \int_0^\infty x^2 (e^x - 1)^{-1} dx. \qquad (6.19)$$

Since the value of the integral is approximately 7.2, the final result is

$$N_{\text{mean}} = 181 \frac{V k_B^3 T^3}{h^3 c^3}. \qquad (6.20)$$

Comparing this result with the expression for the specific heat at constant volume, it is possible to see that C_V is proportional to the product $N_{\text{mean}} k_B$.

6.2.1   *The relation with the wave picture*

All the preceding results have been obtained using the particle picture of the photon gas. In a closed volume, the equivalent of a single photon is a standing wave. The problem is to count the number of standing waves with frequencies between ω and $\omega + d\omega$. Once the result is obtained, one can calculate the energy of the ensemble of the standing waves, using first the equivalence between frequency and energy ($E = \hbar\omega$), and secondly the statistic weight given by $\exp(\hbar\omega/(k_B T)) - 1$.

6.3   Light Emission and Absorption of Solids; Kirchhoff's Law

We come back to the basic phenomenon of emission of light by solids when they are heated. One considers two aspects: the emission and the absorption. We shall see how they are related.

The emission power function of a heated body $B(\lambda, T)$ gives the energy of the radiation emitted by a small area of the solid surface, in a small wavelength interval of $d\lambda$, per area per time, in a small solid angle $d\Omega$, as

$$dE_{em} = B(\lambda, T)(\cos\theta)\, d\lambda\, d\Omega. \qquad (6.21)$$

The angle θ is the angle between the normal of the surface and the direction of the emission. $B(\lambda, T)$ is a function of the wavelength and the temperature; it reflects also the nature of the body. The $\cos\theta$ appears in (6.21) because one considers the emission from a small surface element dA, which is seen as $(\cos\theta)\, dA$ in the θ direction. The emission power, expressed as a function of the wavelength and the temperature, has the dimension of energy per time per volume.

It is possible also to define in the same manner the radiation absorbed by a small area of the surface of the body. It concerns only the radiation which is absorbed and not the radiation reflected or diffused by the surface. The function $M(\lambda, T)$ gives the fraction of the radiation absorbed by the body.

Kirchhoff's law states that the ratio $B(\lambda, T)/M(\lambda, T)$ depends only on λ and T and does not depend on the properties of the body. This ratio is a universal function of λ and T. The exact expression is

$$\frac{B(\lambda, T)}{M(\lambda, T)} = \frac{2\pi hc^2}{\lambda^5}\left[\exp\left(\frac{hc}{\lambda k_B T}\right) - 1\right]^{-1}; \qquad (6.22a)$$

it can be written as

$$\frac{B(\lambda, T)}{M(\lambda, T)} = \frac{c}{2\pi}\frac{8\pi V hc}{\lambda^5}\left[\exp\left(\frac{hc}{\lambda k_B T}\right) - 1\right]^{-1}. \qquad (6.22b)$$

Kirchhoff's law can be applied to the inner walls of the cavity with a photon gas. So it not astonishing that one can write (6.22b) as

$$\frac{B(\lambda, T)}{M(\lambda, T)} = \frac{c}{2\pi}K(\lambda). \qquad (6.23)$$

Recall that $K(\lambda)$ is the energy density per wavelength, of the photon gas (see (6.11b)).

We shall not give a derivation of this law but only make mention an intuitive argument. Consider the emission in a narrow range of wavelengths, from the walls of the cavity where the photon gas is established, its intensity depends on the function $B(\lambda, T)$. There is also absorption by the walls, of energy from the photon gas with energy density $K(\lambda)$. Since these two processes must be equal at the equilibrium (the emitted energy flux must be equal to the absorbed energy flux), one can understand intuitively the expression (6.23), which states $B(\lambda, T) = M(\lambda, T)(c/2\pi)K(\lambda)$.

6.4 The Black Body Emission

A black body is defined as a body with perfect absorption for all wavelengths. In other words, the function $M(\lambda, T)$ is a constant equal to 1. It results from (6.23) that the emission power of the black body is directly related to the spectrum of the photon gas:

$$B(\lambda, T) = \frac{c}{2\pi}K(\lambda).$$

It explains the equivalence between the two phenomena. The study of light emission by solids, which are not black bodies, is beyond the frame of this book.

6.5 The Properties of Photon Gas are Independent of the Shape and the Material of the Cavity

Suppose that the properties of the photon gas inside a cavity, in particular its energy, are not only dependent on the volume and the temperature but also on the shape and the material of the body in which the cavity was made.

One considers two cavities with the same volume and the same shape at the same temperature but the two bodies are made with different materials. The two bodies with their cavities can be seen as a closed system at the temperature T. The energies of the two cavities are different because of the differences in the materials. Conversely, if the energies of the two photon gases were equal, their temperature would be different.

Now one connects the two cavities by means of a tube with a very small volume, giving possibility for the photons of both cavities to diffuse from one cavity to the other. After some time the energies of the two cavities are equal, and consequently their temperatures are different, but one has apparition of a temperature difference in a closed system without making work. This is in contradiction with the second principle of Thermodynamics which states exactly the opposite. This demonstrates that the initial hypothesis is not correct and that the properties of a photon gas in a cavity are independent of the material.

Similarly, following the same argument, one can show that the shape of the cavity is not important.

Chapter 7

Atomic Vibration in Solids: Phonons

7.1 Atomic Vibration in Solids

In this chapter, the thermodynamic properties of solids are studied. This problem is very complicated since atoms move in three dimensions and their motion is collective. An atom cannot move without influencing its neighboring atoms. In the framework of this book one can only present a simplified picture which keeps the essentials of the atomic properties. In particular the specific heat of solids is well described by the model exposed in this chapter, the Debye model. We have already noted that the first model of atomic motions in solids was proposed by Einstein: each atom is seen as independent of its neighbors. Consequently all the atoms move as harmonic oscillators with the same characteristic frequency.

We shall describe briefly the properties of the motion of atoms in a crystalline solid. In a crystal the positions of atoms in the space has some regularities such that the distances between equilibrium positions of atoms are well defined.

At temperature zero all the atoms are placed at their equilibrium positions. When the temperature increases they perform small oscillations around their equilibrium positions for which their amplitudes increases with the temperature. The first important point is that the atomic vibrations are the superposition of several harmonic motions given by standing waves. We take the simple example of atoms in one dimension along a line of length L. The displacements of the atoms can be perpendicular to the line (transverse motion) or along the line (longitudinal motion). A standing wave is the superposition of two propagating waves in reverse directions and opposite amplitudes. For example the first wave is given by $A\sin(\omega t - kx)$ where A is the amplitude, ω is the frequency multiplied by 2π, t is the time, k is the wave vector $k = 2\pi/\lambda$, λ is the wavelength and x is the position along the line. The second wave is given by $-A\sin(\omega t + kx)$. Their sum is

$$A[\sin(\omega t - kx) - \sin(\omega t + kx)] = A\cos(\omega t)\sin(kx).$$

The motion of each point along the line is a harmonic motion, but the amplitude depends on the position. In particular if for $x = 0$ and $x = L$ the amplitude is null when the product $kL = n\pi(n = 1, 2, 3, \ldots)$ or $\lambda = 2L/n$: this is the case of a standing wave. In other words the length of the line is a multiple of a half of the wavelength.

We shall develop this simple model of a linear solid made of a line of N atoms separated the distance a, such that the length[1] of the atoms line is $L = Na$. The atoms are connected by springs with constant B. At $T = 0$ they are at rest and the distance between two consecutive atoms is a. When the temperature increases the atoms perform harmonic motions under the influence of the springs. We suppose, for the sake of simplicity, that the atomic motions are longitudinal. One looks for the relation between the frequency ω and the wave vector k. When one atom (labeled n when n is between 1 and N) is out of its equilibrium position, two forces act on this atom, one from the atom labeled $n-1$ and the other from the atom labeled $n+1$. If we call u_n the displacement of the atom n from its equilibrium position, these forces are proportional to the differences between the displacements of two neighboring atoms, i.e. to $u_n - u_{n-1}$ and $u_n - u_{n+1}$. The equation of motion of the chain is

$$m\frac{d^2 u_n}{dt^2} = -B(u_n - u_{n-1}) + B(u_{n+1} - u_n), \qquad (7.1)$$

---

[1]Rigorously the length of the chain is $a(N-1)$, but since $N \gg 1$ one can take $L = Na$.

where m is the mass of one atom. We look for wave solutions given by

$$u_n = D \sin(\omega t - kna) \tag{7.2}$$

when the quantity na gives the position of the atom n. To check if (7.2) is solution of (7.1), one has to put it in the differential equation (7.1) and to verify for what condition that the two sides of (7.1) are equal. This will give a relation between ω and k.

In place of using the solution (7.2) we shall consider the function $u_n = D \exp[i(\omega t - kna)]$ for which the imaginary part is (7.2). Introducing the exponential in (7.1) we obtain the condition we look for.[2] This gives

$$-m\omega^2 D \exp(i\omega t) \exp(-ikna)$$
$$= BD \ \exp(i\omega t)\{-2\exp(-ikna) + \exp[-ik(n+1)a]$$
$$+ \exp[-ik(n-1)a]\}. \tag{7.3a}$$

We write again this equation as follows

$$-m\omega^2 D \exp(i\omega t) \exp(-ikna)$$
$$= BD \exp(i\omega t) \exp(-ikna)[-2 + \exp(-ika) + \exp(ika)], \tag{7.3b}$$

since $\exp[-ik(n+1)a] = \exp(-ikna) \exp(-ika)$.

Using the relation $e^{ix} + e^{-ix} = 2\cos(x)$ gives $[\exp(-ika) + \exp(ika)] = 2 \cos(ka)$. Introducing in (7.3b) and simplifying, one gets

$$m\omega^2 = 2B[1 - \cos(ka)]. \tag{7.4}$$

Having $\cos(x) = \cos(-x)$, the expression (7.4) is also good for other solutions of (7.1) given by $u_n = D \exp[i(\omega t + kna)]$ propagating in the reverse direction as we need for standing waves solutions. Using the relation

$$\cos(x) = 1 - 2\sin^2\left(\frac{x}{2}\right),$$

one has the relationship between ω and k as

$$\omega = 2\left(\frac{B}{m}\right)^{1/2} \sin\left(\frac{ka}{2}\right). \tag{7.5}$$

---

[2] At the end of the chapter we give the solution of (7.1) by means of trigonometric functions.

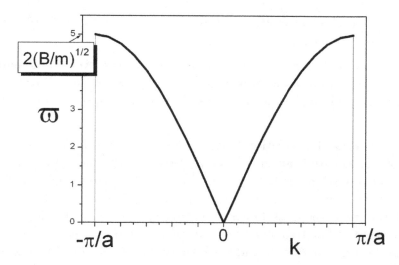

Figure 7.1. Variation of the frequency with the wave vector for a linear atom chain.

In Fig. 7.1, we show the graph of the frequency ω versus the wave vector for the two possible directions of the wave propagation, corresponding to positive k and to negative k. We show this relationship as a continuum, but there only N possible frequencies (see below). From Fig. 7.1 and (7.5) one can deduce the following important characteristics. First there is a maximum frequency given by $\omega_M = 2(B/m)^{1/2}$. Secondly corresponding to this maximum frequency, there is a maximum wave vector (in absolute value) $k = \pi/a$ and a minimum wave length $\lambda = 2\pi/k = 2a$. Thirdly, in writing that the standing wave condition is that the length L is a multiple of a half of the wave length, i.e. $L = d\lambda/2$ where d is an integer, the maximum d (corresponding to the smallest wavelength) is equal to N. In others words, there are N possible frequencies for the chain and they are called the normal modes of the chain. Finally, we note also that only for small wave vectors, the relation between ω and k is linear, using the approximation $\sin x \approx x$ for small x.

To resume, the harmonic function $D\sin(\omega t - kna)$ is solution of the propagating waves in the atom chain with ω related to k by (7.5) and k given by $k = \pm d(\pi/(Na))(d = 1, 2, 3, \ldots, N)$.

Now we shall generalize the above results for the case of three dimension solids.

Property A. Since the atomic motion may be transverse or longitudinal, the motion of an atom is the sum of three harmonic motions, two corresponding to the transverse waves and one to the longitudinal wave.

Property B. The relation between $\dot\omega$ and k is not linear and since the wave velocity is given by the derivative $d\omega/dk$, the velocity is not independent of the frequency as in the case of electromagnetic waves. Only for the small frequencies the relationship between the frequency and the wave vector is linear, resulting in independent velocity. In this range of low frequencies and large wavelengths one has $\omega = kv$ (v is the wave velocity), and since there are two types of waves one has two wave velocities: v_T (transverse wave) and v_L (longitudinal wave). These small frequencies correspond to the small wave vectors and large wavelengths which are sound waves. Effectively the sound frequencies are relatively small (say less than 20 000 Hz) and since the sound velocity is typically several thousands of meters per second in solids, it results that the wave length is much larger than the atomic distances which are of order of several Angstroms. The wave vector is very small. Thus v_T and v_L are the sound velocities in the solid.

Property C. There is an upper limit for the frequencies of the waves as one saw above. This frequency is not easy to calculate and is different for the longitudinal and the transverse waves. Below we shall present the method of the Debye model to get an approximate value.

Property D. There is a limited number of possible frequencies for the atomic vibrations. Since the N atoms of the solid correspond to $3N$ oscillators, the total umber of possible frequencies is also $3N$, in analogy with the case of the linear atomic chain.

Property E. In quantum mechanics the equivalence between waves and particles was established, and we shall use it in the case of the atomic vibrations. The motion of the system of N atoms is described by the superposition of the normal modes which are standing waves.

A standing wave is the wave description of a particle in a restricted volume, as we saw in the case of the photon gas. Similarly to the case of the electromagnetic waves where the particles are photons, one introduces a new particle, the phonon.

7.2 The Properties of Phonons

In complete analogy with the photons, the phonons are bosons and their number is not fixed. Consequently the chemical potential is null. The well known relation between frequency and energy is again $E = h\nu = h\omega/2\pi$, where h is the Planck constant. The relation between the linear momentum with the wave vector is $p = hk/2\pi$ as in the case of the photons.

However there are also important differences. First there are two types of phonons, the transverse phonons corresponding to the transverse waves and the longitudinal phonons corresponding to the longitudinal waves. We recall that the photons correspond only to the transverse electromagnetic waves. Secondly their energy levels are in a finite number. Since in the wave picture there is a finite number of frequencies, one has the same properties for the energies levels. As seen above this is equivalent to saying that there is a maximum possible energy E_M. Finally the relation between the energy and the momentum is linear only for small energies. In this case one has $E = pv_T$ for the transverse phonons and $E = pv_L$ for the longitudinal phonons.

7.3 The Low Temperature Case

We begin with the calculation of the partition function given in Chapter 3 for the case of bosons with variable number of particles.

$$F = -k_B T \sum_I -\mathrm{Ln}[1 - \exp(-\beta e_i)], \qquad (3.13)$$

or going to the energy continuum

$$F = k_B T \int_0^{E_M} g(E) \,\mathrm{Ln}[1 - \exp(-\beta E)] \, dE. \qquad (7.6)$$

In (7.5) the limits of the integral are from 0 to the maximum energy E_M. One has to calculate the function $g(E)$ from the basic expression of $g_p(p)$ given in Chapter 4,

$$g_p(p)\,dp = \frac{s_z(4\pi V p^2\,dp)}{h^3} \tag{4.4c}$$

However, this is not possible because we do not know the relation between E and p except for the low energies. This is precisely the case of the low temperatures where the energy levels occupied by the phonons are the lowest levels. One has two expressions for $g(E)$ one for the transverse phonons and one for the longitudinal ones. Using the relation $p = E/v$, one gets from (4.4c):

Longitudinal phonons $g_L(E)\,dE = \dfrac{4\pi V}{h^3(v_L)^3}E^2\,dE$

Transverse phonons $g_T(E)\,dE = 2\dfrac{4\pi V}{h^3(v_T)^3}E^2\,dE$

We added a factor of 2 in the case of the transverse phonons as we did in the case of the photons. Finally the total function, $g(E)$, is

$$g(E) = g_T(E) + g_L(E),$$

$$g(E) = \frac{4\pi V}{h^3}\left[\frac{1}{v_L^3} + \frac{2}{v_T^3}\right]E^2, \tag{7.7}$$

or writing a mean sound velocity v as $3/v^3 = 1/(v_L)^3 + 2/(v_T)^3$, the final expression is

$$g(E) = \frac{4\pi V}{h^3}\frac{3}{v^3}E^2\,dE. \tag{7.8}$$

We can write the free energy (7.6) in the following form

$$F = \frac{12\pi k_B T V}{h^3 v^3}\int_0^{E_M} E^2\,\mathrm{Ln}\left[1 - \exp\left[-\frac{E}{k_B T}\right]\right]\,dE. \tag{7.9}$$

This expression is very similar to the free energy of a photon gas:

$$F = \frac{8\pi k_B T V}{h^3 c^3}\int_0^{\infty} E^2\,\mathrm{Ln}\left[1 - \exp\left[-\frac{E}{k_B T}\right]\right]\,dE. \tag{6.14a}$$

But the important difference stands in the limits of the integral — in (6.14) they are 0 and ∞ but in (7.8) they are 0 and E_M.

To calculate F given by (7.9), we first note that the function of E in the integrand goes very quickly to zero for large E when T is small. Consequently, one can take the upper limit of the integral equal to infinity. Furthermore, we shall use again the trick used above, of multiplying and diving the expression by $(k_B T)^3$ one gets

$$F = \frac{12\pi (k_B T)^4 V}{h^3 v^3}$$
$$\int_0^{E_M/kT} \left(\frac{E}{k_B T}\right)^2 \text{Ln}\left[1 - \exp\left[-\frac{E}{k_B T}\right]\right] d\left(\frac{E}{k_B T}\right), \quad (7.10)$$

with new limits for the integral: 0 and $E_M/(k_B T)$. But since we consider the low temperature case $(E_M \gg kT)$, one can take the limits as 0 and ∞. The calculation is now completely analogous with that of the electromagnetic waves, (cf. 6.14b). Writing $E_M/(k_B T) = x$, (7.10) becomes

$$F = \frac{12\pi (k_B T)^4 V}{h^3 v^3} \int_0^\infty x^2 \text{Ln}[1 - \exp(-x)]\, dx, \quad (7.11)$$

where we recall that the limits of the integral are now 0 and ∞. The value of the integral is $-\pi^4/45$, and the final result is

$$F = -\frac{4}{15} \frac{\pi^5 (k_B T)^4 V}{h^3 v^3}. \quad (7.12)$$

The energy is calculated through the usual formula $E = F - T(\partial F/\partial T)$, and the specific heat at constant volume from $C_V = \partial E/\partial T$. The final result is

$$C_V = \frac{16}{5} \frac{\pi^5 (k_B)^4}{h^3 v^3} V T^3. \quad (7.13)$$

The variation of C_V with the temperature as T^3 is well verified for solids at low temperatures.

7.4 The High Temperature Case

We begin again from the free energy (7.6)

$$F = k_B T \int_0^{E_M} g(E) \text{Ln}[1 - \exp(-\beta E)]\, dE; \quad (7.6)$$

the limits of the integral are 0 and the maximum energy E_M. Since we are considering high temperatures ($k_B T \gg 1$) we can write

$$\exp\left[-\frac{E}{k_B T}\right] \approx 1 - \frac{E}{k_B T}.$$

Introducing the result to (7.6), we get

$$F = k_B T \int_0^{E_M} g(E) \, \mathrm{Ln}\left(\frac{E}{k_B T}\right)$$

$$= k_B T \int_0^{E_M} g(E) \left[\mathrm{Ln}(E) - \mathrm{Ln}(k_B T)\right] dE, \qquad (7.14)$$

or

$$F = k_B T \int_0^{E_M} g(E) \, \mathrm{Ln}(E) \, dE - k_B T \mathrm{Ln}(k_B T) \int_0^{E_M} g(E) \, dE. \quad (7.15)$$

Writing $A = \int_0^{E_M} g(E) \, \mathrm{Ln}(E) \, dE$ and $J = \int_0^{E_M} g(E) \, dE$, we can express F in the following form:

$$F = A k_B T - J k_B T \, \mathrm{Ln}(k_B T). \qquad (7.16)$$

The quantities A and B depend on N and V but are independent of the temperature. We can calculate the energy $E = F - T(\partial F / \partial T)$ and the result is

$$E = J k_B T. \qquad (7.17)$$

The final step is to calculate $J = \int_0^{E_M} g(E) \, dE$. It is exactly the sum of all the possible states, which is the number of the possible frequencies or energies. J is equal to $3N$, as we saw above. Thus we have

$$E = 3 N k_B T, \qquad (7.18)$$

and

$$C_V = 3 N k_B. \qquad (7.19)$$

7.5 The Debye Formula

In the model of Debye it is supposed that the relation between the energy and the momentum is linear up to E_M. This means that expression for $g(E)$ is given by

$$g(E) = \frac{4\pi V}{h^3} \frac{3}{v^3} E^2 \, dE, \qquad (7.8)$$

and it is assumed to be valid up to E_M. Now E_M can be calculated from the relation

$$\int_0^{E_M} g(E)\, dE = 3N. \tag{7.20}$$

From (7.8) one has

$$\int_0^{E_M} g(E)\, dE = \frac{4\pi V}{h^3}\frac{3}{v^3}\int_0^{E_M} E^2\, dE = 3N. \tag{7.21}$$

The integral is equal to $E_M^3/3$. This gives for E_M

$$E_M^3 = \frac{3Nh^3v^3}{4\pi V}. \tag{7.22}$$

Now one can calculate the energy from (3.23),

$$E = \sum_i e_i\{\exp[\beta(e_i - \mu)] \pm 1\}^{-1}. \tag{3.23}$$

Using this formula for the bosons with $\mu = 0$, and passing to the continuum limit, one has

$$E = \int_0^{E_M} Eg(E)\left[\exp\left(\frac{E}{k_BT}\right) - 1\right]^{-1} dE. \tag{7.23}$$

Substituting (7.8) in (7.23) gives

$$E = \frac{4\pi V}{h^3}\frac{3}{v^3}\int_0^{E_M} E^3\left[\exp\left(\frac{E}{k_BT}\right) - 1\right]^{-1} dE. \tag{7.24}$$

We recall that the limits of the integral are 0 and E_M. To calculate this integral we use once again the trick to transform the integral, by multiplying it and then dividing it by $(k_BT)^4$. This gives

$$E = \frac{4\pi V}{h^3}\frac{3}{v^3}(k_BT)^4\int_0^{E_M}\left(\frac{E}{k_BT}\right)^3\left[\exp\left(\frac{E}{k_BT}\right) - 1\right]^{-1} d\left(\frac{E}{k_BT}\right). \tag{7.25}$$

Writing $x = Ek_BT$, the expression for the energy is

$$E = \frac{4\pi V}{h^3}\frac{3}{v^3}(k_BT)^4\int_0^{x_M} x^3[\exp(x) - 1]^{-1} dx. \tag{7.26}$$

The integral is a function of the temperature through the upper limit $x_M = E_M/(k_B T)$ and can be calculated numerically. It is usual to define the Debye temperature T_D by the relation $E_M = k_B T_D$, and the limit x_M becomes equal to T_D/T. It is possible to transform (7.26) in noting that $4\pi V/(h^3 v^3) = 3N/E_M^3$. One gets

$$E = \frac{9N}{E_M^3}(k_B T)^4 \int_0^{T_D/T} x^3 [\exp(x) - 1]^{-1} dx, \qquad (7.27)$$

or

$$E = 9Nk_B T \left(\frac{T}{T_D}\right)^3 \int_0^{T_D/T} x^3 [\exp(x) - 1]^{-1} dx. \qquad (7.28)$$

Finally one sees that

$$E = 3Nk_B T f\left(\frac{T}{T_D}\right), \qquad (7.29)$$

where the function

$$f\left(\frac{T}{T_D}\right) = 3\left(\frac{T}{T_D}\right)^3 \int_0^{T_D/T} x^3 [\exp(x) - 1]^{-1} dx$$

is a universal function of the ratio T/T_D, valid for all materials.

From (7.28) one can get the expression for the specific heat at constant volume

$$C_V = \left(\frac{\partial E}{\partial T}\right)_V,$$

$$C_V = 9Nk_B \left(\frac{T}{T_D}\right)^3 \int_0^{T_D/T} x^4 \exp(x)[\exp(x) - 1]^{-2} \, dx. \qquad (7.30)$$

The passage form (7.28) to (7.30) necessitates some steps in the derivation. We put it at the end of the chapter. We recall that the limits of the integrals in (7.14)–(7.17) are 0 and $x_M = T_D/T$.

C_V must be calculated numerically. We plot $C_V/(Nk_B)$ as a function of T/T_D in Fig. 7.2. It is possible to show that in the low temperature limit one recovers the T^3 behavior of the specific heat. More precisely one has

$$C_V = \frac{12\pi^4}{5} Nk \left(\frac{T}{T_D}\right)^3, \qquad (7.31)$$

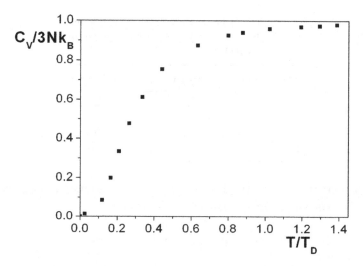

Figure 7.2. Variation of the specific heat of the Debye model as a function of the reduced temperature, T/T_D.

and the measurement of C_v at low temperatures gives a determination of T_D (of course, if the behavior of T^3 is observed).

In the high temperatures limit, the Debye model gives $C_V = 3Nk_B$ as expected. Between these two limits ($T \to 0$ and $T \to \infty$), the Debye model gives a good description of C_v for materials with a structure that is not too complicated.

Another means to check the validity of the Debye model is through the sound velocities. We recall the definition of E_M

$$(E_M)^3 = (k_B T_D)^3 = \frac{3Nh^3 v^3}{4\pi V}, \tag{7.32}$$

and one can relate the Debye temperature to the mean sound velocity v

$$T_D = v\frac{h}{k_B}\left(\frac{3N}{4\pi V}\right)^{1/3}. \tag{7.33}$$

In Table 7.1 we give the values of the Debye temperature obtained from the specific heat and from the sound velocities for some materials. We see that the agreement with theory is very good, taking into account that the Debye model is in fact an approximation.

Table 1.[3] Debye temperature for some
materials. The temperatures are given in
degrees of Kelvin.

Materials	T_D from C_V	T_D from v
NaCl	308	320
KCl	230	246
Ag	225	216
Zn	308	305

7.6 Resolution of the Differential Equation (7.1) by Means of Trigonometric Functions

We recall the differential equation giving the motion of atoms in the
linear atom chain

$$m\frac{d^2u_n}{dt^2} = -B(u_n - u_{n-1}) + B(u_{n+1} - u_n). \tag{7.1}$$

We look for solution with the form $u_n = D\sin(\omega t - kna)$. Introducing into (7.1) gives

$$-m\omega^2\sin(\omega t - kna) = -2B\sin(\omega t - kna) + B\sin[\omega t - k(n-1)a]$$
$$+ B\sin[\omega t - k(n+1)a]. \tag{7.34}$$

One can develop the second and third term in the right side of
(7.34) using the formula

$$\sin(a+b) = \sin(a)\cos(b) + \cos(a)\sin(b),$$

i.e.

$$\sin(\omega t - k(n-1)a) = \sin(\omega t - kna + ka)$$
$$= \sin(\omega t - kna)\cos(ka) + \cos(\omega t - kna)\sin(ka), \tag{7.35a}$$

$$\sin(\omega t - k(n+1)a) = \sin(\omega t - kna - ka)$$
$$= \sin(\omega t - kna)\cos(ka) - \cos(\omega t - kna)\sin(ka). \tag{7.35b}$$

The sum of these two terms is

$$2\sin(\omega t - kna)\cos(ka),$$

---

[3] Adapted from C.Kittel, *Introduction to Solid State Physics* John Wiley and Sons, New York, 1956.

and introducing into (7.34) gives, after some manipulation,

$$m\omega^2 = 2B[1 - \cos(ka)],$$

as above.

7.7 Derivation of the Expression (7.30) Giving C_V in the Debye Model

We begin with the expression of the energy

$$E = 9Nk_BT\left(\frac{T}{T_D}\right)^3 \int_0^{T_D/T} x^3[\exp(x) - 1]^{-1}dx, \tag{7.28}$$

and we divide the two sides by $9T_DNk_B$

$$\frac{E}{9T_DNk_B} = \left(\frac{T}{T_D}\right)^4 \int_0^{T_D/T} x^3[\exp(x) - 1]^{-1}\,dx. \tag{7.36}$$

We consider the group

$$A = \frac{E}{9T_DNk_B} \tag{7.37}$$

and one has

$$\frac{dA}{dT} = \frac{\frac{dE}{dT}}{9T_DNk_B}$$

$$= \frac{C_v}{9T_DNk_B} \tag{7.38}$$

One can write A as a function of $y = T/T_D$ as

$$A = y^4 \int^{1/y} x^3[\exp(x) - 1]^{-1}\,dx \tag{7.39}$$

We derive A relative to y, to get the derivative relative to T, since one has

$$\frac{dA}{dy} = \frac{dA}{dT}\frac{dT}{dy} = T_D\frac{dA}{dT}.$$

We write $f(x) = x^3(\exp x - 1)^{-1}$ such that $A = y^4 \int_9^{1/y} f(x)dx$ and we define the function $F(x)$ by the indefinite integral

$$F(x) = \int x^3[\exp(x) - 1]^{-1}\,dx$$

Consequently $dF/dx = f(x)$. The integral $\int_9^{1/y} x^3(\exp x - 1)^{-1}\,dx$ is equal to $F(1/y) - F(0)$ and A becomes

$$A = y^4 F\left(\frac{1}{y}\right) \quad \text{since } F(0) = 0.$$

(Note that for small x, $f(x)$ behaves as x^2 and $F(x)$ as $x^3/3$.) The derivative dA/dy is

$$\frac{dA}{dy} = 4y^3 F\left(\frac{1}{y}\right) + y^4\left(\frac{-1}{y^2}\right)\left(\frac{dF}{dx}\right)_{1/y}$$

The function $F(x)$ can be calculated by parts in writing $u = [\exp(x) - 1]^{-1}$, $dv = x^3 dx$, $v = x^4/4$ and $du = -\exp(x)[\exp(x) - 1]^{-2}dx$,

$$F(x) = \int u\,dv = uv - \int u\,dv,$$

$$F(x) = x^4[4(\exp(x) - 1)]^{-1} + \frac{1}{4}\int x^4 \exp(x)[\exp(x) - 1]^{-2}dx.$$

$$(7.40)$$

From (7.40) one gets

$$F\left(\frac{1}{y}\right) = \frac{1}{4y^4}\left[\exp\left(\frac{1}{y}\right) - 1\right]^{-1} + \frac{1}{4}\int_9^{1/y} x^4 \exp(x)[\exp(x) - 1]^{-2}dx,$$

$$(7.41)$$

and one has

$$\left(\frac{dF}{dx}\right)_{\frac{1}{y}} = f\left(\frac{1}{y}\right) = \left(\frac{1}{y^3}\right)\left[\exp\left(\frac{1}{y}\right) - 1\right]^{-1}.\qquad (7.42)$$

Putting these expressions (7.41) and (7.42) in (7.39) gives

$$\frac{dA}{dy} = y^3 \int_9^{1/y} x^4 \exp(x)[\exp(x) - 1]^{-2}dx$$

and

$$\frac{dA}{dT} = \frac{T^3}{T_D^4}\int_9^{1/y} x^4 \exp(x)[\exp(x) - 1]^{-2}\,dx,$$

from which one reaches the final formula with the help of (7.37):

$$C_v = 9Nk_B\left(\frac{T}{T_d}\right)^3 \int_0^{T_D/T} x^4 \exp(x)[\exp(x) - 1]^{-2}\,dx.\qquad (7.30)$$

Chapter 8

The Boson Gas at Low Temperature: The Bose–Einstein Condensation

In this chapter, and the following, we consider an ideal gas made of atoms or molecules which are bosons or fermions. By ideal gas, we mean a gas made of particles (atoms or molecules) without interaction with each other, such that the energy of a particle is only the kinetic energy $E = p^2/(2m)$. We have already studied this gas in the classical conditions, i.e. when the difference between bosons and fermions cannot be made. We recall that the gas behave classically if the following condition is obeyed

$$k_B T \gg \frac{h^3}{2\pi m}\left(\frac{N}{V}\right)^{2/3}, \qquad (4.15)$$

that is the temperature is high enough when the gas density is fixed. But if, at constant density, the temperature is lowered, deviations from the classical behavior are noticeable. It is the subject of these chapters to study the behavior of gases at low temperatures. When the temperature is lowered, the negative chemical potential μ increases until it becomes null. This appears at a particular temperature that can be taken as the temperature separating the classical regime from the quantum regime. However the behavior of bosons or fermions is completely different. Sometimes, the quantum regime is called the degenerate state and the classical regime the non-degenerate state.

We begin with the bosons in which the most dramatic change appears at very low temperatures when an important fraction of the atoms is in the lowest state of energy (that we take equal to zero).

8.1 The Chemical Potential

We have observed previously in the chapter on quantum statistics that the properties of the system can be obtained from the knowledge of the chemical potential μ.

To calculate the thermodynamic quantities, one has, following the results of Chapter 3, to determine the chemical potential as a function of the temperature through Eq. (3.22b) applied to the case of the bosons

$$N = \sum_i \{\exp[\beta(e_i - \mu)] - 1\}^{-1}, \tag{8.1}$$

or passing to the energy continuum,

$$N = \int_0^\infty g(E)\{\exp[\beta(E - \mu)] - 1\}^{-1}\, dE, \tag{8.2}$$

where the integration limits are 0 and ∞. The density of states $g(E)\, dE$ is deduced from the expression of the density of states $g(p)\, dp$ obtained in Chapter 4,

$$g_p(p)\, dp = \frac{s_z(4\pi V p^2\, dp)}{h^3}, \tag{4.4}$$

through the relation between the momentum p and the energy E, $E = \frac{p^2}{2m}$ (m is the mass of an atom). One has $p^2 = 2mE$ and $dp = m dE/p = \sqrt{2}m^{3/2}E^{1/2}\, dE$, and inserting in (4.4) one gets (taking s_z equal to 1)

$$g(E)dE = \frac{4\pi V}{h^3}\sqrt{2}m^{3/2}E^{1/2}\, dE. \tag{8.3}$$

Introducing (8.3) in (8.2) gives an expression from which on can deduce the chemical potential

$$N = \frac{4\pi V}{h^3}\sqrt{2}m^{3/2}\int_0^\infty E^{1/2}\{\exp[\beta(E - \mu)] - 1\}^{-1}\, dE, \tag{8.4}$$

or

$$N = \frac{2\pi V (2m)^{3/2}}{h^3} \int_0^\infty E^{1/2} \left\{ \exp\left(\frac{E - \mu}{k_B T}\right) - 1 \right\}^{-1} dE. \quad (8.5)$$

The limits of the integral are 0 and ∞. This is an implicit equation $N/V = F(T, \mu)$, from which one can determine μ as a function of T and the density N/V. We recall also that in the case of bosons the chemical potential must be smaller than the lowest energy; this means that μ must be negative. Furthermore we saw that the derivative $d\mu/dT$ is negative. Lowering the temperature makes the chemical potential smaller in absolute value and we pose the question: what is the limit at low temperature? Surprisingly one can see that Eq. (8.5) has the solution $\mu = 0$ for a finite temperature T_o that one can easily calculate. Effectively, putting $\mu = 0$ and $T = T_o$ in (8.5) gives

$$N = \frac{2\pi (2m)^{3/2} V}{h^3} \int_0^\infty E^{1/2} \left[\exp\left(\frac{E}{k_B T_o}\right) - 1 \right]^{-1} dE. \quad (8.6)$$

After multiplying the right hand side of (8.6) by $(k_B T_o)^{3/2}$, dividing $E^{1/2}$ by $(k_B T_o)^{1/2} dE$ by $k_B T_o$, one gets (after rearranging the factor before the integral)

$$N = V \left[\frac{2\pi m k_B T_o}{h^2} \right]^{3/2} \left(\frac{2}{\sqrt{\pi}} \right) \int_0^\infty x^{1/2} [\exp(x) - 1]^{-1} dx, \quad (8.7)$$

where $x = E/(k_B T_o)$, or

$$\frac{N}{V} = 2.61 \left[\frac{2\pi m \, k_B T_o}{h^2} \right]^{3/2}. \quad (8.9)$$

In (8.9) we replaced the integral by its value. Finally one has

$$T_o = 0.53 \frac{h^2}{(2\pi m \, k_B)} \left(\frac{N}{V} \right)^{2/3}. \quad (8.10)$$

Thus, above T_o, the chemical potential is negative, but below that it is positive. One concludes that something is wrong with the expression (8.2)!

The answer stands in the passage from the series (8.1) to the sum (8.2). In the series, the first term is $[\exp(\frac{-\mu}{k_B T}) - 1]^{-1}$, different from zero, but the function in the integral begins from zero. In fact in passing from the series to the integral, we have eliminated the first term with energy zero. It is just this state which becomes more and more populated when the temperature is lowered.

Thus, below T_o, Eq. (8.2) must be changed and written as follows, putting the first term explicitly in the sum

$$N = \left[\exp\left(\frac{-\mu}{k_B T}\right) - 1\right]^{-1} + \left(\frac{4\pi V}{h^3}\right)\sqrt{2}m^{3/2}$$

$$\times \int_0^\infty E^{1/2}\{\exp[\beta(E - \mu)] - 1\}^{-1}\, dE. \qquad (8.11)$$

One can determine the behavior of μ when the temperature goes to zero. In this limit, N_1 (the number of particles in the lowest energy state) goes to N, and one writes

$$N_1 = \left[\exp\left(\frac{-\mu}{k_B T}\right) - 1\right]^{-1} \to N, \qquad (8.12)$$

or

$$\exp\left[\frac{-\mu}{k_B T}\right] \to \frac{N + 1}{N}. \qquad (8.13)$$

Since N is very large in a macroscopic sample, the ratio $(N+1)/N$ is very near to 1 and this mean that $\mu/(k_B T)$, is very small. In this case one develops the exponential term $\exp(-x) \approx (1-x)$, and (8.13) becomes

$$1 - \frac{\mu}{k_B T} \to \frac{N + 1}{N}, \qquad (8.14)$$

and finally,

$$\mu \to -\frac{k_B T}{N}. \qquad (8.15)$$

The chemical potential is thus very small at very low temperatures and it is a very good approximation to take it equal to zero below and at T_o.

To resume, one writes the general expression (8.11) in the following forms:

(i) For $T > T_o$,

$$N = \frac{2\pi(2m)^{3/2}V}{h^3}\int_0^\infty E^{1/2}\left[\exp\left(\frac{E - \mu}{k_B T - 1}\right)\right]^{-1}\, dE; \qquad (8.5)$$

(ii) For $T = T_o$,

$$N = \frac{2\pi(2m)^{3/2}V}{h^3} \int_0^\infty E^{1/2}\left[\exp\left(\frac{E}{k_B T_o}\right) - 1\right]^{-1} dE;$$

(8.6)

(iii) For $T < T_o$,

$$N = N_1 + \frac{2\pi(2m)^{3/2}V}{h^3} \int_0^\infty E^{1/2}\left[\exp\left(\frac{E}{k_B T}\right) - 1\right]^{-1} dE.$$

(8.16)

The first term N_1 corresponds to the number of particles in the state with energy zero, and the second term corresponds to particles number in all the other states. Remark that in (8.16) the chemical potential is taken equal to zero.

One can get from the preceding formulas (8.16) and (8.7) an expression of N_1 as a function of T_o. Introducing the same procedure as that used above to pass from (8.6) to (8.7), one can write (8.16) as follows:

$$N = N_1 + V\left[\frac{2\pi m\, k_B T}{h^2}\right]^{3/2} \frac{2}{\sqrt{\pi}} \int_0^\infty x^{1/2}[\exp(x) - 1]^{-1} dx.$$

(8.17)

The number of particles in the state with positive energy, $N(e > 0)$ is

$$N(e > 0) = V\left[\frac{2\pi m\, k_B T}{h^2}\right]^{3/2} \frac{2}{\sqrt{\pi}} \int_0^\infty x^{1/2}[\exp(x) - 1]^{-1} dx.$$

(8.18)

One recalls the expression (8.7)

$$N = V\left[\frac{2\pi m\, k_B T_o}{h^2}\right]^{3/2} \frac{2}{\sqrt{\pi}} \int_0^\infty x^{1/2}[\exp(x) - 1]^{-1} dx.$$ (8.7)

From (8.7) one has

$$\frac{2}{\sqrt{\pi}} \int_0^\infty x^{1/2}[\exp(x) - 1]^{-1} dx = \frac{N}{V\left(\frac{2\pi m k_B T_o}{h^2}\right)^{3/2}},$$

and upon insertion of the integral into (8.17) one has $N = N_1 + N(T/T_o)^{3/2}$, and finally

$$\frac{N_1}{N} = 1 - \left(\frac{T}{T_o}\right)^{3/2}. \tag{8.19}$$

The temperature T_o can be seen as the temperature for which the "condensation" of the particles into the lowest state begins. For $T \to 0$, the number of particles in this state goes to N, and for $T \geq T_o$, this number is null. This phenomenon was predicted by Einstein using a method developed by Bose, it is given the name of the Bose–Einstein condensation.

8.2 The Energy, Specific Heat, Free Energy and Entropy

We consider the case $T \leq T_o$, which is simpler because the chemical potential is null. The energy is given by

$$\begin{aligned}
E &= \int_0^\infty E g(E) \left[\exp\left(\frac{E}{k_B T}\right) - 1\right]^{-1} dE \\
&= \int_0^\infty E \left(\frac{4\pi V}{h^3}\right) \sqrt{2} m^{3/2} E^{1/2} \left[\exp\left(\frac{E}{k_B T}\right) - 1\right]^{-1} dE.
\end{aligned} \tag{8.20}$$

One uses again the trick of transforming the integrals, by dividing and multiplying this expression by $(k_B T)^{5/2}$. Rearranging the numerical factor, one has

$$E = V \left[\frac{2\pi m}{h^2}\right]^{3/2} \frac{2}{\sqrt{\pi}} (k_B T)^{5/2} \int_0^\infty x^{3/2} [\exp(x) - 1]^{-1} dx. \tag{8.21}$$

One recalls the expression giving N, (8.7),

$$N = V \left[\frac{2\pi m \, k_B T_o}{h^2}\right]^{3/2} \frac{2}{\sqrt{\pi}} \int_0^\infty x^{1/2} [\exp(x) - 1]^{-1} dx. \tag{8.7}$$

Dividing (8.21) by (8.7), term by term, gives

$$\frac{E}{N} = \frac{T^{5/2} \int_0^\infty x^{3/2} [\exp(x) - 1]^{-1} dx}{(T_o)^{3.2} \int_0^\infty x^{1/2} [\exp(x) - 1]^{-1} dx}. \tag{8.22}$$

The ratio of the two integrals is about 0.77. Thus one gets, finally

$$E = 0.77 N k_B \frac{T^{5/2}}{(T_o)^{3.2}} \tag{8.23}$$

It is possible to express E as a function of V and T, introducing in (8.23) the values of T_o given by (8.10),

$$E = \left(\frac{2\pi m}{h^2}\right)^{3/2} V (k_B T)^{5/2} = A V T^{5/2}, \tag{8.24}$$

where A is equal to $(2\pi m/h^2)^{3/2}(k_B)^{5/2}$. Now one can calculate the specific heat at constant volume C_V, with the entropy and the free energy F:

$$C_V = \left(\frac{\partial E}{\partial T}\right)_V = \frac{5}{2} A V T^{3/2}, \tag{8.25}$$

$$S = \int \frac{C_V}{T} dT = \frac{5}{3} A V T^{3/2}, \tag{8.26}$$

$$F = E - TS = -\frac{2}{3} A V T^{5/2} = \frac{-2}{3} E. \tag{8.27}$$

From (8.27), one deduces the pressure

$$P = -\left(\frac{\partial F}{\partial V}\right)_T = \frac{2}{3} A T^{5/2}. \tag{8.28}$$

One remarks the very particular behavior of the boson gas in the condensation regime. The energy, free energy and entropy do not depend on the number of particles. It is merely the consequence of the fact that the chemical potential is zero. The pressure is independent of the volume.

Immediately above T_o, the calculations appear cumbersome and we do not present them. At temperatures well above T_o, the gas behaves as a classical ideal gas with energy $(3/2)\, N k_B T$.

We shall give a qualitative picture of the specific heat C_V. For $T \leq T_o$, one comes back to the expression (8.23) of the energy, to calculate again the specific heat at constant volume,

$$C_V = \left(\frac{\partial E}{\partial T}\right)_V = 1.925 N k_B \left(\frac{T}{T_o}\right)^{3/2}.$$

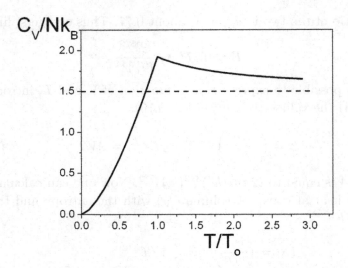

Figure 8.1.   Variation of the specific heat with the reduced temperature, showing the peak in the slope at $T = T_o$.

Well above T_o one has the classical value $C_V = \frac{3}{2}Nk_B$. However, for $T = T_o$, C_V is equal to $1.925Nk_B$, larger than the classical value. It is possible to show that the specific heat is continuous at T_o, and that for $T > T_o$, C_V begins to decrease from $1.925\ Nk_B$ to $3/2\ Nk_B$. In Fig. 8.1 one shows schematically the variation of C_V with T.

8.3 Experimental Verfication

The Bose–Einstein condensation was predicted in 1925 and at the time there was no experimental evidence of the phenomenon. The first proposition for a possible experimental realization was made by London in 1938 about the properties of liquid helium. This element is a gaseous at room temperature and it remains liquid even at very low temperatures. Particular properties appear at the temperature of $2.3°$K, when it losses its viscosity. This state is called a superfluid state. The idea of London was to consider this transition as a Bose–Einstein condensation. If one applies the formula (8.10) to calculate the Bose–Einstein condensation, one obtains T_o at about $3.4°$K. It is not exactly the good temperature but it is not too far. To understand the proposal of London, one considers the two fluid models proposed later by Tisza. In this model, liquid helium in its superfluid state is compound of two fluids. In the first the atoms form a normal liquid and in the other the atoms form the superfluid liquid. The atoms in this superfluid state are those condensate in the lowest energy level as in the Bose–Einstein condensation. However, the application of the Bose–Einstein condensation to helium presents several difficulties, since the theory considers atoms in their gaseous state when they do not have any interaction between them, in contrast to liquid helium.

The problem is an experimental realization of the Bose–Einstein condensation is as follows. From the expression (8.10) of T_o, it is possible to see that the condensation can occur only at low temperatures when all the elements are solid (with the exception of helium which remains liquid). If the number of atoms is reduced to a small number (for example, 10^6 atoms or less) it is possible to kept them in gaseous state. However the condensation temperature drops to very low (below 10^{-6} °K). The technical difficulties are to get a gas of small number of atoms at a very low temperature. Only very recently, solution to this problem was found, and the Bose–Einstein condensation was observed. These are very sophisticated techniques, first to cool the atoms (like Rubidium atoms) by laser and evaporation, and secondly to keep the group of atoms far from the wall of the container (to avoid heating of the atoms) by magnetic interaction. The atoms in the condensation state are in their lowest kinetic energy state, and consequently are almost stationary.

Chapter 9

The Gas of Fermions: Electrons in Metals and in Semiconductors

In this chapter, we consider a gas of electrons in two particular situations. In the first, we present a simple model of metals described by a gas of N electrons in a box of volume V. The model is applied to describe the properties of metals, since in a metal a part of the electrons are not located near the nuclei but are free to move in all the volume of the metal sample. It is a surprisingly good approximation for some metals.

It is a surprise, since in the theoretical derivation of the properties of the gas, the electrons are seen as independent particles when in the real metals the free electrons are subject to their coulomb interaction. We shall not present explanation for the success of the model but only make a mention of it.

In the second part of the chapter, we consider the case of semiconductors which are characterized by the absence of free electron when the temperature is zero. However when the temperature is increased, some electrons become free and form a gas of electrons.

9.1 Free Electrons in a Box

9.1.1 *The Fermi–Dirac function*

Our goal is to find the energy of the N electrons as a function of T, N and V. We recall the formula of the energy we got in,

Chapter 3, (3.23):

$$E = \sum_i e_i \{\exp[\beta(e_i - \mu)] \pm 1\}^{-1}$$

In the present case, one takes the + sign and (3.23) becomes

$$E = \sum_i e_i \{\exp[\beta(e_i - \mu)] + 1\}^{-1}, \tag{9.1}$$

or in passing to the continuum,

$$E = \int_0^\infty E g(E) \{\exp[\beta(E_i - \mu)] + 1\}^{-1} \, dE, \tag{9.2}$$

where $g(E)$ is the density of states for a massive particle with kinetic energy $p^2/2m$. The function $g(E)$ was already calculated in the preceding chapter and it is:

$$g(E) \, dE = \frac{8\pi V}{h^3} \sqrt{2} m^{3/2} E^{1/2} \, dE. \tag{9.3}$$

In (9.3) we introduced the factor 2, to take into account the spin of the electrons, which is $1/2$. Thus the energy is given by

$$E = \frac{8\pi V}{h^3} \sqrt{2} m^{3/2} \int_0^\infty E^{3/2} \{\exp[\beta(E_i - \mu)] + 1\}^{-1} \, dE. \tag{9.4}$$

In (9.2) and (9.3) the limits of the integrals are from 0 to ∞. One sees from (9.3) that, in principle, to calculate E one needs to calculate the chemical potential μ. For historical reason, it is frequently called the Fermi level. Sometimes the nomenclature is restricted to the value of μ at $T = 0$ but in the present book, it will be called always "Fermi level", whatever the temperature.

The determination of μ is made through the general expression of N, (3.22b),

$$N = \sum_i \{\exp[\beta(e_i - \mu)] \pm 1\}^{-1},$$

or in our case,

$$N = \frac{8\pi V}{h^3} \sqrt{2} m^{3/2} \int_0^\infty E^{1/2} \{\exp[\beta(E - \mu)] + 1\}^{-1} \, dE. \tag{9.5}$$

In order to be able to calculate μ from (9.4) and E from (9.3), a study of the function

$$f_{FD} = \{\exp[\beta(E - \mu)] + 1\}^{-1} \tag{9.6}$$

is useful. This function is called the Fermi–Dirac function. We recall that this function gives the probability that a state with energy E is occupied by an electron.

First we consider the FD function for $T = 0$ when the Fermi level is $\mu_0 > 0$. If the energy E is smaller than $\mu_0, \exp[\beta(E - \mu_0)]$ goes to 0 if $T \to 0 (\beta \to \infty)$ since $E - \mu_0 < 0$. In this case f_{FD} goes to 1. Now if E is larger than μ_0, $\exp[\beta(E - \mu_0)] \to \infty$ as $T \to 0 (\beta \to \infty)$, since $E - \mu_0 > 0$. In this case, the function f_{FD} is null. Consequently the f_{FD} function is a step function passing from the value 1 to the value 0 when the variable E crosses the value $E = \mu_0$. The function is shown in Fig. 9.1.

We have already seen in Chapter 3 that the occupied levels are the first N levels. This means that at $T = 0$, the Fermi level is equal to the highest energy occupied by the last particle.

When the temperature increases (but it not too high), the f_{FD} function takes the shape indicated in Fig. 9.2 with $\mu > 0$. For $E \ll \mu$ the function is equal to 1, but near μ it decreases steeply to 0. For $E = \mu$ the function is equal to $1/2$.

If now one increases strongly the temperature, so that the Fermi level becomes negative, the f_{FD} can be approximated by an exponential: $f_{FD} = \exp[-\beta(E - \mu)]$ since $\exp[\beta(E - \mu)] \gg 1$. (See Fig. 9.3.)

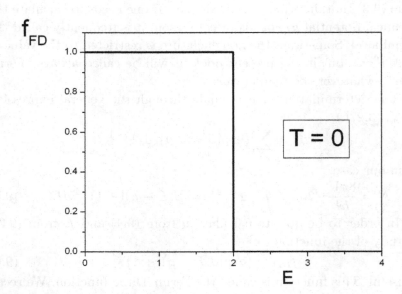

Figure 9.1.   The Fermi–Dirac function for $T = 0$. The energies are in electron-volts and the Fermi level is taken equal to $2\,\text{eV}$.

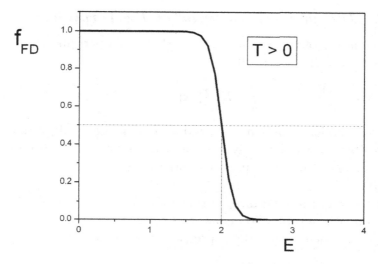

Figure 9.2.   The Fermi–Dirac function for $T > 0$. The energies are indicated in eV. In this figure $k_B T$ is equal to 0.08 eV. Since k_B is equal to 8.6210^{-5} eV/$^\circ$K, the temperature is 929 $^\circ$K.

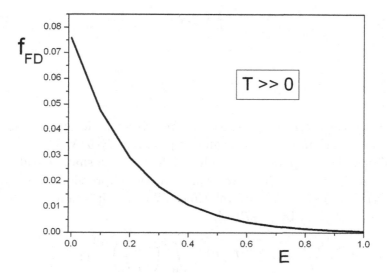

Figure 9.3.   The Fermi–Dirac function when the Fermi level is negative. It is equal to -0.5 eV and the temperature is 2309 $^\circ$K. The energies are indicated in eV. Note that the Fermi–Dirac function is very low and is even lower if T is smaller.

9.1.2 *The chemical potential or the Fermi level*

The variation of the Fermi level μ with the temperature can be calculated with the help of (9.5)

$$N = \frac{8\pi V}{h^3}\sqrt{2}m^{3/2}\int_0^\infty E^{1/2}\{\exp[\beta(E-\mu)]+1\}^{-1}\,dE. \qquad (9.5)$$

It is an implicit function of T and μ. Further, it is clear that it is impossible to get an analytic expression giving $\mu(T)$. However at least for one value of T it is possible to get an exact value, i.e. for $T = 0$.

We saw in the preceding section that at $T = 0$, the Fermi–Dirac function is a step function equal to 1 for $E < \mu_0$ and equal to 0 for $E > \mu_0$. In such a case, the expression (9.5) reduces to

$$N = \frac{8\pi V}{h^3}\sqrt{2}m^{3/2}\int_0^\mu E^{1/2}\,dE. \qquad (9.7)$$

The integral is equal to $(2/3)(\mu_0)^{3/2}$. This gives for N

$$N = \frac{8\pi V}{h^3}\sqrt{2}m^{3/2}\frac{2}{3}\mu_0^{3/2}. \qquad (9.8)$$

From which one gets

$$\mu_0 = \left(\frac{N}{V}\right)^{2/3}\frac{h^2}{2m}\left(\frac{3}{8\pi}\right)^{2/3}. \qquad (9.9)$$

One takes a typical value for the free electrons in metals, $N/V = 5 10^{28}\,\mathrm{m}^{-3}$ and finds for μ_0 a value approximately $5\,\mathrm{eV}$.

When the temperature is different from 0, we shall consider two different limits. For that, we transform the expression (9.5), using the trick already used several times. We multiply and divide the right side of (9.4) by $(k_B T)^{3/2}$,

$$N = \frac{8\pi V}{h^3}\sqrt{2}m^{3/2}(k_B T)^{3/2}\int_0^\infty \left(\frac{E}{k_B T}\right)^{1/2}d\left(\frac{E}{k_B T}\right)$$

$$\times\left\{\exp\left[\frac{E-\mu}{k_B T}\right]+1\right\}^{-1}, \qquad (9.10)$$

or

$$N/V = \frac{4}{\sqrt{\pi}} \left[\frac{2\pi m \, k_B T}{h^2} \right]^{3/2}$$

$$\times \int_0^\infty x^{1/2} \left[\exp\left(\frac{-\mu}{k_B T} \right) \exp(x) + 1 \right]^{-1} dx, \quad (9.11)$$

when we write $x = \frac{E}{k_B T}$. The two cases we consider are those with

$$\frac{N}{V} \ll \frac{4}{\sqrt{\pi}} \left[\frac{2\pi m \, k_B T}{h^2} \right]^{3/2} \qquad (9.12)$$

and

$$\frac{N}{V} \gg \frac{4}{\sqrt{\pi}} \left[\frac{2\pi m \, k_B T}{h^2} \right]^{3/2}. \qquad (9.13)$$

We begin with the first condition (9.12), which means that the electron density is small. When it is fulfilled, the integral of (9.11) (that we call I) reduces to

$$I = \int_0^\infty x^{1/2} \left[\exp\left(-\frac{\mu}{k_B T} \right) \exp(x) + 1 \right]^{-1} dx \ll 1.$$

This occurs when μ is negative and $\exp(-\mu k_B T) \gg 1$. In other words the situation is classical or non-degenerate. In such a case the integral becomes

$$I = \exp\left(\frac{\mu}{k_B T} \right) \int_0^\infty x^{1/2} \exp(-x) \, dx = \exp\left(\frac{\mu}{k_B T} \right) \frac{\sqrt{\pi}}{2}.$$

$$(9.14)$$

Inserting the value of the integral in (9.11) gives

$$\frac{N}{V} = 2 \left[\frac{2\pi m \, k_B T}{h^2} \right]^{3/2} \exp\left(\frac{\mu}{k_B T} \right), \qquad (9.15)$$

and one gets

$$\mu = k_B T \, \text{Ln}\left(\frac{N}{V} \right) - k_B T \, \text{Ln}\left\{ 2 \left[\frac{2\pi m \, k_B T}{h^2} \right]^{3/2} \right\}. \qquad (9.16)$$

We have already found all these results. The inequality (9.12) is the condition for the classical case, and is equivalent to the condition

(4.16). Next, the expression (9.16) for the chemical potential is the equivalent of the expression (3.29) when one inserts the complete expression (4.7) for Z_1. The only difference stands in the factor 2 appearing in (9.16) since we took into account the electron spin.

Now we consider the case given by the inequality (9.11) which corresponds to a large electron density and relatively low temperatures. It is the degenerate case. One needs to use the complete expressions (9.4) or (9.5). As mentioned above, it is not possible to get an analytic expression $\mu(T)$ giving the variation of the Fermi level with the temperature. However a numerical calculation is feasible and we performed it in order to give an illustration of the variations of μ with T.

We chose $\mu_0 = 2\,\text{eV}$ which corresponds to a electron density of about $5 \times 10^{26}\,\text{m}^{-3}$. We verify that the inequality (9.13) in calculating the quantity $(4/\sqrt{\pi})[(2\pi m k_B T)/h^2]^{3/2}$ for $= 500\,°\text{K}$. We get $3.4 \times 10^{25}\,\text{m}^{-3}$, i.e. the inequality (9.13) is just satisfied. In such case the complete curve (Fig. 9.4) shows that the Fermi level is practically constant up to this temperature. Important changes in μ appear only

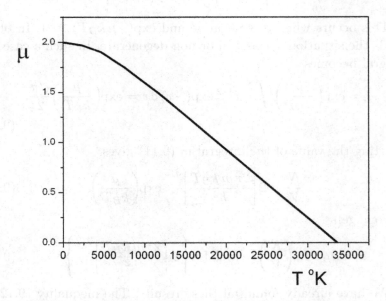

Figure 9.4.   Variation of the Fermi level with the temperature.

at very high temperatures, where μ decreases until it becomes null for $T = 33540°$K. Of course no element can stand this temperature and we show the complete curve for the sake of the completeness. The conclusion is that for large densities and "reasonable" temperatures the Fermi level is constant.

We can now confess that it is possible to have (after hard calculation work) an expression for the Fermi level at low temperature. It is

$$\mu = \mu_0 - \frac{\frac{\pi^2}{12}(k_B T)^2}{\mu_0}. \tag{9.17}$$

The reader can verify (for example for $T = 500°$K) that the relative change in the Fermi level $(\mu - \mu_0)/\mu_0$ is miniature such that the conclusion of the preceding paragraph is correct.

9.1.3 *The energy*

Case 1: $T = 0$.

The calculation of the energy E at $T = 0$ is not difficult due to the simple form of the Fermi–Dirac function. We recall the general expression for E, (9.4),

$$E = \frac{8\pi V}{h^3}\sqrt{2}m^{3/2}\int_0^\infty E^{3/2}\{\exp[\beta(E_i - \mu)] + 1\}^{-1}\, dE.$$

The expression of E becomes

$$E_0 = \frac{8\pi V}{h^3}\sqrt{2}m^{3/2}\int_0^\mu E^{3/2}\, dE. \tag{9.18}$$

The integral is equal to $(2/5)\, \mu_0^{5/2}$, and one has

$$E_0 = \frac{8\pi V}{h^3}\sqrt{2}m^{3/2}\frac{2}{5}\mu_0^{5/2}. \tag{9.19}$$

We recall a preceding expression (9.8),

$$N = \frac{8\pi V}{h^3}\sqrt{2}m^{3/2}\frac{2}{3}\mu_0^{3.2},$$

and divide side by side the expression (9.19) by (9.8), to get

$$\frac{E_0}{N} = \frac{3}{5}\mu_0 = 0.0727\left(\frac{N}{V}\right)^{2/3}\left(\frac{h^2}{2m}\right). \tag{9.20}$$

Case 2: $T \neq 0$.

The determination of E for $T \neq 0$ is much more difficult and before we derive an expression valid at low temperatures, we need to recall two properties relative to definite integrals, which are not well known.

Property 1. The permutation of the limits of the integral changes its sign

$$\int_a^b f(x)\,dx = -\int_b^a f(x)\,dx.$$

Property 2. It is applicable if the function in the integrand is defined for positive and negative values of the variable

$$\int_{-a}^0 f(x)\,dx = \int_0^a f(-x)\,dx.$$

We begin with the expression (9.4) that we write in the following form,

$$E = D\int_0^\infty h(E)\left\{\exp\left[\frac{E-\mu}{k_BT}\right]+1\right\}^{-1} dE, \tag{9.21}$$

where $D = (8\pi V/h^3)\sqrt{2}m^{3/2}$ and $h(E) = E^{3/2}$.

The Fermi level appears in (9.21). In principle it is temperature dependent, and in the temperature range we consider it constant and equal to its value at $T = 0$, as explained in the preceding section.

We have to calculate the integral:

$$J = \int_0^\infty h(E)\left\{\exp\left[\frac{E-\mu}{k_BT}\right]+1\right\}^{-1} dE \tag{9.22}$$

in the limit of low temperatures. As a first step one makes a change in the variable $x = (E-\mu)/(k_BT)$ and has $E = \mu + k_BTx$, $dE = k_BT\,dx$ and the limits are now $-\mu/(k_BT)$ and ∞.

The integral J becomes

$$J = k_B T \int_{-\frac{\mu}{kT}}^{\infty} h(\mu + k_B T x)[\exp(x) + 1]^{-1}\, dx. \qquad (9.23)$$

In the second step the integral is written as the sum of two integrals, $J = J_1 + J_2$, where

$$J_1 = k_B T \int_{-\frac{\mu}{kT}}^{0} h(\mu + k_B T x)[\exp(x) + 1]^{-1}\, dx, \qquad (9.24a)$$

$$J_2 = k_B T \int_{0}^{\infty} h(\mu + k_B T x)[\exp(x) + 1]^{-1}\, dx. \qquad (9.24b)$$

Using Property 2 above the integral J_1 becomes

$$J_1 = k_B T \int_{0}^{\frac{\mu}{kT}} h(\mu - k_B T x)[\exp(-x) + 1]^{-1}\, dx. \qquad (9.25)$$

Using the identity

$$[\exp(-x) + 1]^{-1} = 1 - [\exp(x) + 1]^{-1},$$

and inserting it in (9.25), one sees that J_1 can be written as the sum of two integrals J_1' and J_1'':

$$J_1' = k_B T \int_{0}^{\frac{\mu}{kT}} h(\mu - k_B T x)\, dx, \qquad (9.26)$$

$$J_1'' = -k_B T \int_{0}^{\frac{\mu}{kT}} h(\mu - k_B T x)[\exp(x) + 1]^{-1}\, dx. \qquad (9.27)$$

Now we make in J_1' the following change of variable x: $y = \mu - k_B T x$, $dx = -dy/(k_B T)$ and the limits become $y = \mu$ and $y = 0$. Consequently,

$$J_1' = -k_B T \int_{\mu}^{0} h(y)\, dy/(k_B T) = \int_{o}^{\mu} h(y)\, dy. \qquad (9.28)$$

The second equality is due to Property 1 above.

The upper limit of the integral J_1'' is $\mu/(k_B T)$ which is very large when T is very small. Thus we take it infinite. We can give to the integrals J_2 and J_1'' the same limits $[0, \infty]$ and put them together in one sum

$$J''' = k_B T \int_{0}^{\infty} [h(\mu + k_B T x) - h(\mu - k_B T x)][\exp(x) + 1]^{-1}\, dx. \qquad (9.29)$$

It is an important result which can be used if the function $h(y)$ has a simple form like a polynomial. In such a case it is possible to calculate the integral since the values of integrals of the type $\int_0^\infty x^n [\exp(x) + 1]^{-1}$ are knowm (when n is an integer). But in the present case $h(y) = y^{3/2}$ and the integral (9.29) cannot be calculated.

We shall develop the function $h(y)$ appearing in the integrand of J''' in series around the value μ. It is justified as long as $k_B T x = \varepsilon$ is much smaller than μ, and case if this is not the (when x goes to infinite), $\exp(x)$ becomes large enough to make the integrand small and negligible.

One has

$$h(\mu - \varepsilon) = h(\mu) - \varepsilon \left(\frac{dh}{dy}\right)\Big|_\mu + \frac{\varepsilon^2 \left(\frac{d^2 h}{dy^2}\right)_\mu}{2} + \cdots,$$

$$h(\mu + \varepsilon) = h(\mu) + \varepsilon \left(\frac{dh}{dy}\right)\Big|_\mu + \frac{\varepsilon^2 \left(\frac{d^2 h}{dy^2}\right)_\mu}{2} + \cdots,$$

and

$$h(\mu + \varepsilon) - h(\mu - \varepsilon) = 2\varepsilon \left(\frac{dh}{dy}\right)_\mu.$$

Finally the integral J''' is

$$J''' = (k_B T)^2 \left(\frac{dh}{dy}\right)_\mu \int_0^\infty x[\exp(x) + 1]^{-1}\, dx, \qquad (9.30)$$

the integral $\int_0^\infty x[\exp(x) + 1]^{-1}\, dx$ is equal to $\pi^2/12$.
The final result is

$$J = J_1 + J_2 = J_1' + J_1'' + J_2,$$
$$J = J_1' + J'''.$$

Putting the value of J_1' (9.28) and that of J''' (9.30) gives

$$J = \int_0^\mu h(y)\, dy + \frac{\pi^2}{6} (k_B T)^2 \left(\frac{dh}{dy}\right)_\mu. \qquad (9.31)$$

We return back to the initial expression of the energy and introduce the integral J in (9.21):

$$E = DJ = \frac{8\pi V}{h^3} \sqrt{2} m^{3/2} \left[\int_0^\mu h(y)\, dy + \frac{\pi^2}{6} (k_B T)^2 \left(\frac{dh}{dy}\right)\Big|_\mu \right].$$

$$(9.32)$$

Recalling that $h(y) = y^{3/2}$ and $(dh/dy)|_\mu = 3/2\ \mu^{1/2}$,

$$E = \frac{8\pi V}{h^3}\sqrt{2}m^{3/2}\int_o^\mu y^{3/2}\,dy$$

$$+ \frac{8\pi V}{h^3}\sqrt{2}m^{3/2}\frac{\pi^2}{6}(k_BT)^2\frac{3}{2}\mu^{1/2}. \tag{9.33}$$

The first term is the energy for $T = 0$ (see expression (9.18) when the integrand y replaces E), and the second term is due to the contribution of temperature.

The quantity $(8\pi V/h^3)\sqrt{2}m^{3/2}$ appears in the relation between N and μ_0,

$$N = \frac{8\pi V}{h^3}\sqrt{2}m^{3/2}\left(\frac{2}{3}\right)\mu_0^{3/2}. \tag{9.8}$$

One deduces that $(8\pi V/h^3)\sqrt{2}m^{3/2} = (3N)/(2\mu_0^{3/2})$. Introducing it in (9.31) gives the final expression[1] for E in low temperatures:

$$E = E_0 + \frac{\pi^2}{4}N\frac{(k_BT)^2}{\mu_0}, \tag{9.34a}$$

with $E_0 = (3/5)N\mu_0$. Introducing the value of μ_0 in (9.34a) gives

$$E = E_0 + 10.17N\left(\frac{V}{N}\right)^{2/3}(k_BT)^2\left(\frac{2m}{h^2}\right). \tag{9.34b}$$

9.1.4 *The specifc heat*

From the expression of the energy (9.34a) one can calculate the specific heat at constant volume, $C_V = (\partial E/\partial T)_V$,

$$C_V = \frac{\pi^2}{2}\frac{N(k_B)^2T}{\mu_0}. \tag{9.35}$$

It is linear with the temperature.

In several metals this relation is verified at very low temperature. The total specific heat of a metal is compound of the contribution

---

[1]In several textbooks the expression for the energy is derived with the help of the method due to Sommerfeld. The method used here is due to Landau and Lifshitz.

of the atoms and that of the free electrons. However the first contribution is much larger than the electronic contribution except at low temperature when the atomic contribution is very low. This appears because it varies with the temperature as T^3 and is much lower than the electronic part which varies as T.

It is possible to write the specific heat in different forms. One can introduce the Fermi temperature T_F, defined by $k_B T_F = \mu_0$ or $T_F = \mu_0/k_B$. Substituting in (9.33) gives

$$C_V = \frac{\pi^2}{2} k_B N \frac{T}{T_F}. \tag{9.36}$$

Finally we give another expression for C_V. We recall the expression of the number of electrons

$$N = \frac{8\pi V}{h^3} \sqrt{2} m^{3/2} \int_0^\infty E^{1/2} \{\exp[\beta(E-\mu)] + 1\}^{-1} \, dE, \tag{9.5}$$

which can be written as $N = \int_0^\infty n(E) \, dE$. The function $n(E)$ can be interpreted as giving the number of electrons in the vicinity of the energy E. For $E = \mu_0$ one has

$$n(\mu_0) = \left(\frac{8\pi V}{h^3}\right) \sqrt{2} m^{3/2} \mu_0^{1/2}. \tag{9.37}$$

This gives

$$\mu_0^{1/2} = n(\mu_0) \left[\frac{8\pi V}{h^3} \sqrt{2} m^{3/2}\right]^{-1}. \tag{9.38}$$

Above we found that $(8\pi V/h^3)\sqrt{2} m^{3/2} = (3N)/(2\mu_0^{3/2})$, and putting it in (9.38) gives $n(\mu_0) = (3N)/(2\mu_0)$ or $\mu_0 = (3N)/[2n(\mu_0)]$. Inserting this expression of μ_0 in (9.34a),

$$E = E(0) + \frac{\pi^2}{6} (k_B T)^2 n(\mu_0). \tag{9.39}$$

One gets for C_V

$$C_V = \frac{\pi^2}{3} n(\mu_0) \, k_B^2 T. \tag{9.40}$$

The interest of this last formula stands in the possibility to extract $n(\mu_0)$, the number of electrons with energy equal to the Fermi level, from a measure of the electronic specific heat.

9.1.5 *Applications to metals*

We shall give a simplified picture of the energies of electrons in solids and particularly in metals. The energy levels of electrons in an isolated atom are a series of discrete values but when the atoms form a solid, the possible energies for electrons are grouped in bands. The origin of the bands is the fact that the atoms are not isolated but interact in order to form a solid. It results that the density of states for electrons in solids is compound of several distinct regions called "bands." In Figs. 9.5 and 9.6 we give the two possible situations for solids at temperature zero. In the first (Fig. 9.5), the first bands are completely occupied by the electrons and the last band is empty. In Fig. 9.5 we show an example of three occupied bands and one empty band. It is possible to show that a filled band cannot sustain an electric current, and thus the material is an insulator.

In the other situation, the last band is not completely filled. We show in Fig. 9.6 such an example of two bands completely filled when the third is only partially filled. This band is called the conduction

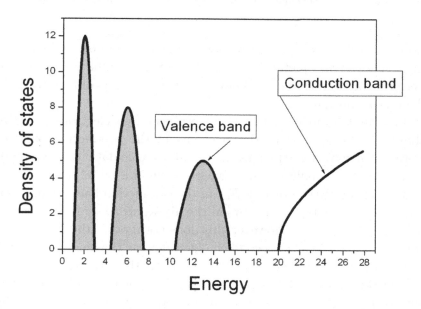

Figure 9.5.   Schematic picture of the electronic bands in an insulator.

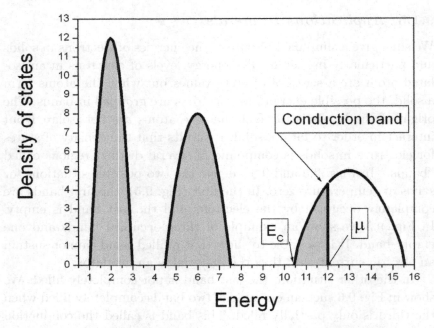

Figure 9.6.   Schematic picture of the electronic bands in a metal.

band since the electrons in this band can sustain an electric current. This material is a metal. The electrons of the conduction band are free electrons which are not bonded to specific atoms differently from the case of insulators.

In a first approximation, the picture of free electrons we gave above can be applied to the electrons in the conduction band of metals, but with two changes. In the above calculations it was supposed that the smallest energy is zero when in a metal, the smallest energy is that at the bottom, E_C, of the conduction band. Thus each time that the energy E appears in the above expressions, we have to replace it by $E - E_C$. However this does not modify the expressions of the specific heat. The second change concerns the mass of the electron. One of the influences of the atoms on the free electrons is to change their mass, which is now an effective mass. But this effect is not strong, such that in our approximation it is possible to use the electron mass.

9.2 Electrons in Semiconductors

Semiconductors are a particular class of insulators. Like all insulators, the electrons are bonded to the nucleus such that at zero temperature there are no free electrons. However it is possible to liberate electrons from the atoms, and consequently the insulator can sustain an electric current. In the band picture this means that electrons with energy of the last filled band, called the valence band, have gained enough energy to pass from the valence band to the conduction band. In semiconductors this process takes place by heating the material (and also under influence of light).

Once the electrons are liberated and store their energy in the conduction band, what we said above about the free electrons in metals is applicable concerning their energy minimum (E_C) and their effective mass. There is still an important difference — the Fermi level is below the minimum energy E_C of the electrons when for metals it is in the conduction band. Consequently the electrons do not obey the Fermi–Dirac statistics, but rather the Maxwell–Boltzmann statistics. This can be understood in considering Fig. 9.5. Since at $T = 0$ the conduction bamd is empty and the valence band is filled, from the property of the Fermi–Dirac function, one sees that the Fermi level must be between E_C, the bottom of the conduction band, and E_V, the top of the valence band, that is $E_V < \mu < E_C$.

We want to answer the following question: What is the number of electrons in the conduction band when the material is heated? It is important in order to understand electric conduction in the semiconductors which depends strongly on the temperature. It is not the case of the electrons in metals since their number does not depend on the temperature.[2]

To calculate the number of electrons in the conduction band, one needs to know the density of states — for such we use the "parabolic

---

[2]It does not mean that their electric conduction is completely independent of the temperature, since the electric conduction is not only a function of the number of free electrons but also of their ability to move under influence of the electric field. This last property depends slightly on T.

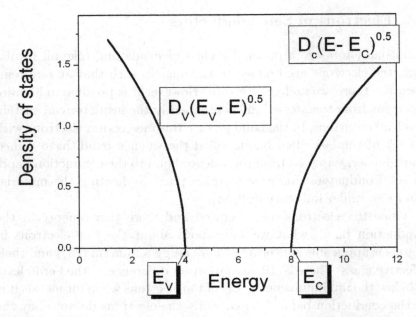

Figure 9.7.   Parabolic approximation of the bands of a semiconductor.

approximation." We assume that the conduction band can be approximate, in the vicinity of E_C, by the following expression:

$$g_c(E) = D_C(E - E_C)^{1/2}, \qquad (9.41)$$

in analogy with the case of free electrons in a box (see (9.3)). We
complete the analogy in writing $D_C = (8\pi V/h^3)\sqrt{2}m_e^{3/2}$ where m_e is
the effective mass of an electron (see Fig. 9.7). The effective electron
mass is defined through D_C. It is often smaller than the electron
mass.

The number of electrons in the conduction band is now

$$N_e = \frac{8\pi V}{h^3}\sqrt{2}m_e^{3/2} \int_{E_C}^{\infty} (E - E_C)^{1/2}\{\exp[\beta(E - \mu)] + 1\}^{-1}\, dE.$$

The limits of the integral are chosen from E_C to ∞. The lower
limit is self-evident but the upper one is incorrect since the parabolic
approximation is valid only for values of E near E_C. However the
Fermi–Dirac function is a rapidly decreasing function such that for
large values of E the integrand is very small and does not contribute

significantly to the integral. Since the Fermi level μ is smaller than E_C the Fermi–Dirac function reduces to an exponential function, and one has

$$N_e = \frac{8\pi V}{h^3} \sqrt{2} m_e^{3/2} \int_{E_C}^{\infty} (E - E_C)^{1/2} \exp[-\beta(E - \mu)]\, dE. \quad (9.42)$$

Now we deal with the electrons in the valence band. In this band there is a lack of electrons equal to the number of electrons in the conduction band. Instead considering the electrons themselves in the valence band, we consider the number of states which are not occupied by the electrons. If the probability for a state to be occupied by an electron is given by the Fermi–Dirac function f_{FD}, the probability that a state is not occupied is $1 - f_{FD}$. The number of unoccupied states is equal to the density of states in the valence band multiplied by $1 - f_{FD}$.

The unoccupied states of the valence band receive an intuitive description. When an electron is bonded to an atom, it moves in a restricted region of the space near the atom. If now this electron is liberated and moves freely, this unoccupied place may receive an electron from the neighboring atoms. In leaving the atom, this new electron creates a new place which can be filled by another electron. Thus the appearance of a free electron brings about the motion of electrons from an unoccupied place to another. These wandering places correspond to the unoccupied states of the valence bands, and are called "holes".

We use again the "parabolic approximation" for the valence band, as shown in Fig. 9.7, and write

$$g_v(E) = D_v(E_V - E)^{1/2} \quad (9.43)$$

and $D_v = (8\pi V/h^3)\sqrt{2} m_h^{3/2}$, where one defines m_h as the effective mass of a hole.

The number of holes is given by

$$N_V = \frac{8\pi V}{h^3} \sqrt{2} m_h^{3/2} \int_{-\infty}^{E_V} (E_V - E)^{1/2}(1 - f_{FD})\, dE, \quad (9.44)$$

where

$$1 - f_{FD} = 1 - \{\exp[\beta(E - \mu)] + 1\}^{-1} = \{\exp[-\beta(E - \mu)] + 1\}^{-1}.$$

Since $E < \mu$, one has $\exp[-\beta(E - \mu)] \gg 1$, and $1 - f_{FD}$ becomes $\exp[\beta(E - \mu)]$. Inserting into (9.44) gives

$$N_V = \frac{8\pi V}{h^3} \sqrt{2} m_h^{3/2} \int_{-\infty}^{E_V} (E_V - E)^{1/2} \exp[-\beta(\mu - E)] \, dE.$$

$$(9.45)$$

Now we are able to calculate the Fermi level as a function of the temperature and the number of electrons in the conduction band, or equivalently, the number of holes in the valence band. We shall make some transformations to the expressions (9.42) and (9.45) to perform it.

First we write in (9.42), $E - \mu = E - E_C - (\mu - E_C)$,

$$\beta(E - E_C) = \frac{E - E_C}{k_B T} = y, \quad dE = (k_B T) \, dy.$$

The expression (9.42) is now

$$N_e = \frac{8\pi V}{h^3} \sqrt{2} (k_B T)^{3/2} m_e^{3/2} \exp[\beta(\mu - E_C)] \int_0^\infty y^{1/2} \exp(-y) \, dy.$$

$$(9.46)$$

In (9.45) we write

$$\mu - E = \mu - E_V + (E_V - E),$$

$$\beta(E_V - E) = \frac{E_V - E}{k_B T} = y, \quad dE = -\frac{k_B T}{dy},$$

and

$$N_V = \frac{8\pi V}{h^3} \sqrt{2} (k_B T)^{3/2} m_h^{3/2} \exp[-\beta(\mu - E_V)]$$

$$\times \int_0^\infty y^{1/2} \exp(-y) \, dy. \qquad (9.47)$$

In writing (9.47) we used Property 1 above, concerning definite integrals. Now we have two equations to find the two unknown quantities: $N_C = N_V$ and μ. The value of the integral appearing in (9.46) and (9.47) is $\sqrt{\pi}/2$. First we multiply (9.46) and (9.47) side by side and find

$$N_C N_V = \frac{32\pi^3 V^2}{h^6}(k_B T)^3 (m_e m_h)^{3/2} \exp[-\beta(E_C - E_V)],$$

and finally

$$N_C = N_V = (N_C N_V)^{1/2} = 2V \frac{2\pi k_B T}{h^2}^{3/2} (m_e m_h)^{3/4}$$

$$\times \exp\left[\frac{-\beta(E_C - E_V)}{2}\right]. \tag{9.48}$$

One sees that the number of electrons in the conduction band and that of the holes in the valence band are strongly dependent on the temperature. In particular it is a function of the difference $E_C - E_V$, which is called the band gap.

To get the Fermi level, one writes, using the equlity of right sides of (9.46) and (9.47). One obtains

$$\mu = \frac{E_C + E_V}{2} + \frac{3}{4}k_B T \operatorname{Ln}\left(\frac{m_h}{m_e}\right). \tag{9.49}$$

For $T = 0$ the Fermi level is exactly at middle distance, from the bottom of the conduction band and the top of the valence band. Since $m_h > m_e$ frequently, the Fermi level is an increasing function of T.

All we describe about the properties of the semiconductor concerns the case where there is no impurity. It is the intrinsic regime. However, if it has impurities, even in a relatively small number, it may introduce electrons and holes. When their numbers from the impurities are higher than the electrons and holes from the material itself, one speaks about the extrinsic regime.

A final word about the holes as missing places of electrons. In the absence of a net electric field, the jumping electrons and the missing places are moving at random. But under influence of an electric field, the electrons move preferentially in the opposite direction of the field. Consequently the missing places move in the direction of the field. They behave as if there were particles with a positive electric charge and with the effective mass of the holes.

Chapter 10

A History of Statistical Mechanics

10.1 Thermodynamics and Statistical Mechanics Before Maxwell and Bolztmann

It is impossible to write a history of statistical mechanics without reference to thermodynamics. These fields are two aspects of the more general field of thermal physics which includes the microscopic and macroscopic sides. However, I shall try to emphasize the important steps which permitted an understanding of the development of statistical mechanics. It is also difficult in a limited space of one chapter (a complete book is necessary) to quote all the scientists who contributed to the advances in the field.

The first precursor of statistical mechanics is probably the mathematician Bernoulli who wrote a treatise of hydrodynamics in 1738. He gave a picture of a gas made of balls bouncing on the walls of the container. He calculated the pressure on the walls by the motion of the ball and derived Boyle's law of the constancy of the product of pressure and volume at constant temperature. He showed also that this product is proportional to the kinetic energy of the balls. One has to note that it is supposed that all the balls have the same velocity. This assumption is adopted by numerous scientists and the work of Mawell described below departs from this view. The work of Bernoulli was forgotten until 1859.

Thermodynamics begins really at the middle of the eightieth century when the properties of gas began to be discovered (Boyle's law and Gay-Lussac's law). At this time the concept of ideal gas was not known and it was thought that all gases have the same properties. The use of heat to produce work began to be realized in steam engines. The concepts of specific heat and of latent heat are developed. Like in all fields of science, experimental and theoretical works appeared side by side. At the beginning of the nineteenth century there is a new area with the formulation of the important concept of energy (Young, 1806), the study the heat propagation (Fourier, 1807), the thermal properties of solids (Dulong and Petit, 1819) and gases (Delaroch and Bérard, 1812), and the problem of understanding the mechanism of the heat engine in order to maximize the work obtained from heat. The contribution of Sadi Carnot in 1824 concerning the transformation of heat in work is fundamental since the basic principles of a cycling engine are explained. He showed that a fraction of the heat that the engine receives from the heat reservoir is expelled outside without being transformed to work.

In the middle of the nineteenth century the basis of thermodynamics was established with the two laws of thermodynamics. In particular the second law received several formulations. One of the most important formulations is based on the concept of entropy introduced by Clausius[1] in 1865. Formally it is proposed that exists a state function of a system defined by the integral taken between two equilibrium states $S = \int dQ/T$, in which dQ is a small reversible amount of heat. It was a mysterious concept related to irreversibility. It was first related to the lack of symmetry concerning the transformation of work in heat and the contrary. It is possible to transform some work completely to heat, but the converse is not true as shown by Carnot. The most impressive aspect is the principle that in an isolated system any irreversible process (or any process taking place spontaneously) results in an increase of entropy.

---

[1]It is usual to mention the entropy by the letter S. The letter was used first by Clausius. There was speculation, that maybe it was done in honor of Sadi Carnot.

In parallel of the development of thermodynamics several theoretical works were done on the theory of gas by Herapath (1812), Laplace (1824), on the interaction between particles by Waterston (1845), Joule (1851) and Clausius (1857). The topic was not in the meanstream of interest for the majority of scientists. One important result of these works is that the mean kinetic energy of a molecule of gas is proportional to the temperature, or more exactly $(1/2)mv_m^2 = (3/2)(R/N_A)(273 + T)$, where N_A is Avogadro's number, R is the gas constant and T is given in centigrade degrees.

10.2 The Kinetic Theory of Maxwell

The name of Maxwell is well known by the fundamental equations of electromagnetism. Elsewhere he worked also in the kinetic theory of gas. In 1860 he published on the determination of the velocity distribution of an ideal gas.

The goal of this work is to find out what proportion of N molecules in a gas has a velocity between v and $v + dv$. One begins by the unknown function $f(v_x)$ which gives the number of molecules with velocity in the x direction between v_x and $v_x + dv_x$, such that their number is $Nf(v_x)\,dv_x$. In other directions y and z, one has similar expressions for the number of molecules with velocity between v_y and $v_y + dv_y$, $Nf(v_y)\,dv_y$, and in the z direction $Nf(v_z)\,dv_z$, since the gas is isotropic. Now for the number of molecules with velocities, in respective directions between v_x and $v_x + dv_x$., v_y and $v_y + dv_y$, v_z and $v_z + dv_z$, one has $Nf(v_x)f(v_y)f(v_z)\,dv_x\,dv_x\,dv_z$ since the velocities in the different directions are independent. But since there is no correlation between the velocities, this number of molecules must depend only the total velocity $(v_x^2 + v_y^2 + v_z^2)^{1/2}$, or

$$f(v_x)f(v_y)f(v_z) = F(v_x^2 + v_y^2 + v_z^2),$$

where F is also an unknown function. It is possible to show that such equality is possible for an exponential function obeying $f(v) = A \exp(\pm Bv^2)$. Maxwell chose the minus sign, since when v increases, the number of molecules must decrease. The final result is that the number of molecules with velocity between v and $v + dv$ is

$$NA^3 \exp(-Bv^2)\, 4\pi v^2\, dv,$$

where $4\pi v^2\, dv$ is the element of volume in the velocity space between the spheres with radius v and $v + dv$. It remains to find the two constants A and B. This is done using the following data

$$N \int_0^\infty A^3 \exp(-Bv^2)\, 4\pi v^2\, dv = N,$$

since the total sum is equal to the number of molecules. And the mean value of v^2 which is known to be equal to $3RT/(Nm)$ is given by

$$\int_0^\infty v^2\, A^3 \exp(-Bv^2)\, 4\pi v^2\, dv = \frac{3RT}{Nm}$$

The final result[2] is

$$f(v) = 4\pi \left(\frac{m}{2\pi k_B T}\right)^{3/2} v^2 \exp\left(-\frac{E}{k_B T}\right).$$

The importance of this work is evaluated from the picture of gases, in which is proposed that each molecule has only a certain probability to have a particular velocity, since the expression $A^3 \exp(-Bv^2)\, 4\pi v^2\, dv$ is seen as a probability. It is the first introduction of probability to thermal physics. One cannot know the velocity of an individual molecule, but only the probability that it has a given velocity. This step is conceptually significant since it paves the way to the development of statistical mechanics.

10.3 Boltzmann and Irreversibility

In the second half of the nineteenth century, an explanation of the existence of irreversibility by means of the laws of mechanics seemed to be impossible. The reason is inlaid within the equations of the mechanics, which are reversible with a change of the time from t to $-t$. We can take a very simple example of throwing a stone into the

---

[2]We write $f(v)$ in the modern terminology. The Boltzmann constant was not known at the time of Maxwell.

air with a velocity striking an angle with the vertical direction. After a parabolic trajectory, the stone reaches the ground. Imagine if it were possible to reverse the time after the throw, we shall retain the same trajectory as the first case, but with a change in the directions of the positions and of the velocities. The process of throwing is thus reversible. Applying this argument to the motions of molecules in a gas shows that the gas will never show irreversibility.

The proposal of Boltzmann to understand irreversibility can be formulated as follows. The state of a system is given by the positions and the velocities of all the components, say N molecules, by the knowledge of $6N$ quantities, 3 for the position of every molecule and 3 for its linear moment, if one sees these molecules as points. One can define a phase–space with $6N$ dimensions, where a "point" in this space represents a microscopic state of the system. It will move continuously in this space. For the sake of the simplicity, this phase–space is divided into cells in which the representative point is located.[3]

The basic idea of Boltzmann is that the different states of the systems have different probabilities to take place. The macroscopic equilibrium state has the largest probability. More precisely, the equilibrium macroscopic state corresponds to the largest number of cell points representing the system.

That a macroscopic state has a certain probability does not exclude the possibility for other states different from the equilibrium state to exist, but that their probability is marginal such that it is not possible to observe them. A simple example of irreversibility is a drop of black ink falling into a water container. Immediately after the fall, one can distinguish colored and transparent regions of the water. A while later, the water takes a uniform gray color, in which one does not see states with separation of ink and water. Such states are not forbidden by the laws of mechanics, but they have a marginal probability to appear, while the state of complete mixture has the largest probability and is the equilibrium state. One

---

[3]We shall not discuss the problem of the size of a cell and the problem if it contains one or more representative points.

sees the qualitative analogy with entropy. In the example of the ink and water, the entropy of the system (ink + water) increases until its largest value is reached at the equilibrium state. A direct relation must exist between the entropy and the probability of the state of equilibrium.

It is the fundamental proposal of Boltzmann (1877) to relate first, the probability of occurrence of a state to the number of cells in the phase–space and secondly, this probability to the entropy. The probability is merely proportional to the number W of points corresponding to a macroscopic state or the number of combinations realizing this state. The entropy is proportional to the logarithm of W or $S = k_B \operatorname{Ln} W$. The constant k_B is the famous Boltzmann constant. In equilibrium W, and consequently S, must be maximum.

This proposal was not accepted immediately by many scientists and the debate about the Boltzmann formula was intense. Echoes of the discussions reverberate long after publication of the paper of Boltzmann.

The importance of this proposal must be emphasized, since it is the first time that an interpretation of this mysterious quantity called entropy is given, a quantity which can only increase in an isolated system. This gives the possibility to relate microscopic and macroscopic quantities. Statistical mechanics was thus born.

Very often, particularly in the popular literature, the entropy is related to disorder. Microscopic systems are seen as disordered since the molecules have various and varying states of both position and velocity. Frequently one says that W and S are measures of disorder, but in the formulation one forgets the link with irreversibility and the macroscopic aspect of the entropy.

10.4 Gibbs, the Father of Statistical Mechanics

The contribution of Gibbs to statistical mechanics can be found in a short treatise published in 1902, a year before his death: *Elementary Principles in Statistical Mechanics*. This *book* had a decisive influence of the development of statistical mechanics, term that Gibbs himself coined. The subtitle indicates its program: *the rational foundation of thermodynamics.*

The goal of Gibbs is to develop rigorously what he calls a "broader view" of mechanics. He adopts the point of view of Maxwell in leaving out the possibility to follow in time the evolution of a mechanic system. He imagines an ensemble of identical systems, varying over the configurations[4] and the velocities, he looks for the law giving the number of systems "which fall within any infinitesimal limits of configuration and velocity." It is clear that Gibbs knew the work of Boltzmann since he mentioned him on his preface. He used also largely the concept of phase–space to describe states of a system.

Gibbs introduces the three ensembles: the canonical, the microcanonical and also the grand ensemble for system compounded of several kinds of molecules. In contrary to the method adopted in this *book*, he defines the canonical ensemble by a linear relation between the logarithm, of the probability of finding a given state, and the energy: $\log P = (\psi - \varepsilon)/\theta$ (ε is the energy, ψ and θ being constant) or $P = \exp[(\psi - \varepsilon)/\theta]$. This choice is justified only by its importance. Gibbs deduces all the properties of the canonical ensemble from this basic definition. He concludes that θ is related to the temperature and P to the entropy.

This small *book* gives us all the methods which are now standard in the study of thermal physics.

10.5 Planck and Einstein: Quantum Theory and Statistics

We put in parallel the work of Planck concerning the black body radiation and that of Einstein on the specific heat of solids. Both gave the correct interpretation of phenomena that classical statistical mechanics was not able to explain, and introduced the basic principle of quantum physics: the energy quanta.

---

[4]By configuration Gibbs meant the positions of the particles of the system.

At the end of the nineteenth century the radiation spectrum of a gas of photon was measured in a large range of wavelengths. It was shown that this spectrum as a function of the frequency (or the wavelength) exhibits a maximum. The classical interpretation begins with the equipartition theorem for one oscillator with energy k_BT. To get the radiation spectrum one multiples the energy of one oscillator by the frequency density of states of standing waves $(V/(\pi^2 c^3))\omega^2\, d\omega$ (ω is 2π the frequency ν multiplied by 2π), and one has $[\omega^2 k_BT/(\pi^2 c^3)]\, d\omega$ for the expression the radiation spectrum which is correct only at low frequencies. Thus the classical theory cannot explain the maximum in the spectrum.

The work of Planck was twofold. Examining the experimental results he found empirically the correct formula:

$$K(\nu, T) = [8\pi h\nu^3/c^3][\exp(h\nu/(k_BT)) - 1]^{-1}.$$

The second step was to find the theoretical derivation of this formula which he did in 1900. He took the method of Boltzmann to calculate the entropy, and from the relation between energy and entropy he determined the temperature. The innovation in calculating W (the number of combinations) was to suppose that the energy of one oscillator was discontinuous by steps of $h\nu$ (h was a constant which was named the Planck constant and ν is the oscillator frequency). This gives the possibility to get the formula previously found only through inspection of the experiments. It is not clear if Planck was conscious that his assumption is to open the path towards a new field in physics. It is likely that he was happy because it gave satisfactory explanation. One must also stress that from his work Planck was able to give the first determinations of his constant h and of the Boltzmann constant k_B.

As mentioned above, at the beginning of the nineteenth century, Dulong and Petit made numerous measurements of specific heats of solids until they formulated their law, stating that the specific heat of solids is independent of T and equal to 6 calories per mole (approximately equal to $3R$). This law was thought so general that it was used to determine the atomic weight of some solids. However, there were exceptions. In particular measurements of carbon made in

1833 by Avogadro and in 1840 by de la Rive and Maret showed that, at ambient temperature, the specific heat of carbon is much lower than 6 calories per mole. In 1872 Weber succeeded in measuring the specific heat of diamond at low temperature (down to 20 K), wherein he observed a regular decrease of the specific heat when the temperature is lowered.

The theoretical interpretation of the Dulong–Petit law was given by Boltzmann in 1876 using the equipartition theorem for N three-dimensional harmonic oscillators, from which he obtained the value of $3R/$mole.

In 1906 the puzzle of the specific heat at low temperature was given its first solution by Einstein. He calculated the mean energy E_m of harmonic oscillators by means of the formula of the canonical ensemble

$$E_m = \frac{\int_0^\infty E \exp\left(-\frac{E}{k_B T}\right) dE}{\int_0^\infty \exp\left(-\frac{E}{k_B T}\right) dE},$$

giving $E_m = k_B T$, the classical result. The idea of Einstein was to adopt the assumption of Planck concerning the energy of an oscillator. He supposed that the energy can take only discrete values by steps of $h\nu$. Letting $E = h\nu/(k_B T)$, he retrieves the mean energy

$$E_m = h\nu[\exp\left(\frac{h\nu}{k_B T}\right) - 1]^{-1}.$$

For solids Einstein supposed that all the atoms have the same frequency. This gives for the specific heat

$$C = 3N k_B \left(\frac{h\nu}{k_B T}\right)^2 \exp\left(\frac{h\nu}{k_B T}\right) \left[\exp\left(\frac{h\nu}{k_B T}\right) - 1\right]^{-2}.$$

He compared his formula with data from diamond and found good agreement except at very low temperatures. The assumption of independent motion of atoms is not correct, for which Debye later in 1912 provided a method to calculate the specific heat, taking into account the collective motions of the atoms.

The works of Planck and Einstein were preliminary introductions of quantum ideas to statistical mechanics. Their successes presents strong evidence that a new physics is needed.

10.6 The Method of Bose and the Bose–Einstein Condensation

In 1924 Bose published a short paper in which he proposed a new derivation of the Planck formula. He wrote it first in English but the paper was refused by an English journal *The Philosophical Magazine.* He wrote to Einstein for his opinion and Einstein subsequently translated it into German. Finally the paper appeared in *Zeitschrift für Physik.* The novelty in this paper of Bose was to see the radiation gas as a gas of photons and to apply the relation $p = h\nu/c$. The density of states here was the momentum density of states. He calculated the entropy by the number of combinations, supposing that the particles are undistinguishable.

The method of Bose was adopted by Einstein in considering molecules and not photons. He calculated W, the number of combinations, leaving out the possibility to distinguish between particles. He looked for the maximum $\mathrm{Ln}\,W$ but with constraints, as Bose did, where the number of particles is constant and the energy is also constant. The well known method of Lagrange multipliers permits to determine a maximum (or a minimum) of a function with constraints.

In this way, Einstein was able to get the well known formula, giving the mean number of particles with energy E

$$N_E = [A^{-1}\exp\left(-\frac{E}{k_B T}\right) - 1]^{-1}.$$

One recognizes in this formula the habitual expression if one take the multiplier A equal to $\exp(\mu/(k_B T)$ (μ is the chemical potential). In a paper in 1924 (shortly after the publication of the paper by Bose) Einstein considered only the case $A < 1$ (or $\mu < 0$), i.e. a boson gas at not too high temperature. But later in 1925 he investigated the case $A = 1$ (or $\mu = 0$) and concluded that a growing number of molecules remains in the smallest energy level. Einstein established the condensation of boson gas. The name Bose is associated to Einstein, since it was Einstein who used the method first proposed by Bose. Today it is a standard method which is exposed in several textbooks. In this book we take another way that is more direct.

10.7 The Principle of Pauli and the Statistics of Fermi and Dirac

The principle of Pauli is strictly a quantum effect. Pauli, in 1925, looking for the interpretation of the electronic structure of atoms, proposed that two electrons cannot be found in the same quantum state. At the same time he saw the need for a new quantum number which is the spin.

The relation with statistics was seen independently by Fermi and Dirac in 1926. In fact Fermi was the first who published a short paper in which he gave the expression of the Fermi–Dirac function. He followed a method parallel to that of Einstein for bosons, maximizing the function $\text{Ln}\,W$ with the two constraints fixing the number of electrons and the energy. The determination of W was made in accordance with the Pauli principle.

Dirac found also the Fermi–Dirac function, but he gave a more general view of the statistics in showing that there are two kinds of particles, bosons and fermions. Considering the wave function of an ensemble of particles, he showed that there are two possibilities when one makes a permutation between the quantum states of two particles. Either there is no change of the wave function or there is a change in the sign. In the last case, it is possible to show that two particles cannot be in the same quantum state.

Interestingly, this novelty in statistical mechanics was immediately appreciated and applied. For example, at end of 1926, Ehrenfest and Uhlenbeck published an article on the two statistics giving their names of Bose–Einstein and Pauli–Fermi–Dirac. In 1927 Sommerfeld established the theory of electrons in metal using the Fermi–Dirac statistics.

Finally the relation with the spin (bosons with integer spin and fermions with half-integer spin) was made independently by Fierz in 1939 and by Pauli in 1940.

Thus, by 1940, all the tools to study statistically independent particles were present.

10.8 Modern Developments

To finish this brief history, I want give some words on recent advances. Thermodynamics is not a closed field and there are always new developments but at relatively slow rate. I want only to quote the names of Onsager and Prigogine who studied the thermodynamics of irreversible processes.

In statistical mechanics, emphasis was put on methods for studying interacting particles. The main field of application is condensed matter although there are also applications in other fields like astrophysics. The examples of applications are numerous (semiconductors, superconductivity, magnetism, theory of liquids, transport phenomena, percolation *etc.*). A very active research is phase transitions and critical points with the theory on renormalization (Wilson, 1972).

The final word is for the application of statistical mechanics outside physics. There are many tentative studies to apply its methods to economics (with a new field, econophysics), to geography, to the social sciences (sociology, psychology) and probably still to others.

Exercises

Chapter 1

Exercise 1.1

A system is made of three identical harmonic oscillators in a closed box. The energy of one oscillator is given by

$$E_i = \hbar\omega\left(n_i + \frac{1}{2}\right) \quad \text{when } i = 1, 2, 3.$$

The n_i's are integers equal to $0, 1, 2, 3 \ldots$. The system is prepared such that the total energy is

$$E = \sum E_i = \hbar\omega\left(n + \frac{3}{2}\right) \quad \text{when } n = n_1 + n_2 + n_3.$$

(a) Suppose that $n = 3$, find the entropy of the system.
(b) Suppose that $n = 5$, find the entropy of the system.
(c) Now n is chosen to be an unspecified integer, show that the entropy is given by

$$S = k_B \operatorname{Ln}[(n+1)(n+2)/2].$$

Calculate S as a function of the energy $S(E)$, and the energy as a function of the temperature T. Give the limits of $E(T)$ for low and high temperatures.

(d) Now one considers N identical harmonic oscillators. The total energy is given by

$$E = \sum E_i = \hbar\omega\left(n + \frac{3N}{2}\right) \quad (i = 1, 2, 3, \ldots, N).$$

The numbers N and n are very large numbers such you can use the Stirling formula $\operatorname{Ln} N! = N \operatorname{Ln} N - N$.

Calculate the entropy as function of n, N and the temperature T, also the energy E as a function of T.

Hint: You have to calculate the number of possibilities to put n units, say "particles," in N "boxes" or oscillators. Suppose that we put the n particles on a line and also put $N - 1$ walls between the particles giving a picture of the n particles in N boxes. You have to find now all the combinations of these $n + N - 1$ objects (n particles and $N - 1$ walls) taking into account that the particles are indistinguishable and also the walls.

Exercise 1.2

A closed system is made of two subsystems A and B in thermal contact one with another. In each subsystem there are N particles with different possible energies. The particles in the subsystem A can have two possible energies, 0 and ε; in the subsystem B they can have two possible energies, ε and 2ε. The total energy E, which is the sum of the energies E_A and E_B, is known.

(a) Calculate the energies E_A and E_B of each subsystem. Express your answers with the help of E, N and ε.

   Hint: Find the condition giving the equality of temperature between side A and side B.

(b) Calculate the temperature of the system.

   Hint: Use the Stirling formula.

Chapter 2

Exercise 2.1

Solve Problem 1.2 by means of the partition function.

Exercise 2.2

N particles are in thermal contact with a reservoir at temperature T. The possible energies of the particles are as follows:

Energy $E_1 = 0$ and degeneracy 1 (only one state with energy E_1)
Energy $E_2 = \varepsilon$ and degeneracy 2 (two states with energy E_2)

Energy $E_3 = 2\varepsilon$ and degeneracy 1 (one state with energy E_3)

(a) Calculate the one-particle partition function, the total partition function and the free energy F.
(b) Calculate the energy of these N particles and find their limits as $T \to 0$ and as $T \to \infty$.
(c) Calculate the entropy and the limits as $T \to 0$ and as $T \to \infty$.

Exercise 2.3

N particles have p possible states with energies $E_i (i = 1, \ldots, p)$. Show that in the limit of high temperatures, the entropy tends toward $S(T \to \infty) = Nk_B \operatorname{Ln}(p)$.

Exercise 2.4

An ideal gas is in a cylindrical container of height L and two extreme bases with area A. The atomic mass is m, the linear momentum is p and their number is N. The container and the atoms are in the gravitational field in the z direction of the cylinder axis.

(a) Calculate the one-atom partition function Z_1.
(b) Calculate the energy of the gas.
(c) What is the probability for a particle to be at the height z? What is the number of particles at the height z?

Hint: Use classical mechanics.

Chapter 3

Exercise 3.1

Show that in a system of quantum particles with energies e_i, the total energy is

$$E = \sum e_i \frac{\partial F}{\partial e_i},$$

where F is the Helmholtz free energy.

Exercise 3.2

The possible energies of N quantum particles are $E_1 < E_2 < E_3 < E_4$ with degeneracy $G_i (i = 1, 2, 3, 4)$.

(a) Sketch a qualitative graph of the chemical potential as a function of the temperature in the following cases:

1. The particles are bosons.
2. The particles are fermions with $N = G_1 + G_2$.

Plot the two curves on the same graph.

(b) Given an expression of the energies in the classical limit (Maxwell–Boltzmann statistics), and find the energy at very high temperatures.

Exercise 3.3

(a) Find the relationship between n_1, the mean number of quantum particles in the lowest level (with energy E_1), and the mean number n_i in one energy level higher E_1, for the following cases:

1. Bosons.
2. Fermions.

(b) Calculate n_i as a function of n_1, the temperature T and the difference $E_i - E_1$.
(c) What are the limits of n_i at high temperatures for bosons (and the same for fermions)?
(d) Find the limit of n_i at low temperature for bosons.
(e) Show that for fermions it is not possible to find the limits of n_i at low temperature.

Chapter 4

Exercise 4.1

(a) Give the expression of the energy E of an ideal gas as a function of the entropy S, the volume V and the number of atoms N.
(b) At what condition is the derivative $(\partial E/\partial N)_{s,v}$ negative?
(c) Is this last condition in agreement with the state of the gas as an ideal gas?

Exercise 4.2

(a) Show that the density of states of particles with linear moment p located on a surface of area A (particles in two dimensions, $2D$) is

$$g_2(p)\, dp = (2\pi A)h^{-2}p\, dp.$$

(b) Show that the density of states of particles with linear momentum p located on a line of length L (particles in one dimension, $1D$) is

$$g_1(p)\, dp = \frac{2L}{h}dp.$$

(c) Calculate the energy density of states $g(E)\, dE$ of particles for whish the energy and the momentum are related with $E = Kp^s(s > 0)$ in $3D$, $2D$ and $1D$.

Exercise 4.3

N atoms of an ideal gas with mass m and linear momentum p are located in a cubic box with edge L. A side of the box can absorb atoms and these atoms behave as an ideal gas in two dimensions.

(a) Calculate the partition function of the ideal gas on the absorbing side if one supposes that there are N_S absorbed atoms on this side.

(b) Calculate the grand partition function of the atoms inside the box and the grand partition function of the absorbed atoms.

(c) Give the ratio N_S/N_V:

  1. By means of the partition functions.
  2. By means of the grand partition functions.

Exercise 4.4

A gas of quantum particles has the following properties:

1. The possible energies form a continuum from 0 to infinity ($E \geq 0$).

2. The energy density of states is $g(E)\,dE = A\,E^\alpha\,dE$.

(a) Demonstrate the following relationship:
$$PV = KE,$$
where K is a function of α. (P is the pressure and V the volume.)

Hint: Transform the integral giving $\operatorname{Ln} Z_G$ with integration by parts.

(b) Apply the results to the cases of relativistic particles $E = pc$ and non relativistic massive particles $E = p^2/(2m)$.

Exercise 4.5

An ideal gas of N atoms have their energy related to the momentum as $E = pc$ (c the velocity of light). The volume is V and the temperature T.

(a) Calculate the one-atom partition function.
(b) Calculate the energy and the specific heat at constant volume.
(c) Calculate the equation of state.

Chapter 5

Exercise 5.1

One considers N distinguishable units, with energy forming a continuum form 0 to infinity. To know their thermal properties $E(T)$ and $S(T)$ (S is the entropy), one student proposes to divide the energy in discrete units such the energies are $E_i = K n_i$, where n_i are integers from 0 to infinity, to perform the calculation and finally to make $K \to 0$.

(a) Solve the problem by means of the microcanonical ensemble. Suppose that the energy is $E = Kn$ and find the entropy of the system with the help of E, N and K. Calculate $E(T)$ and find its limit for $K \to 0$.

Hint: See Problem 1.2 and use the Stirling formula.

(b) Solve the problem by means of the partition function.

Exercise 5.2

Find the energy and the entropy as functions of the temperature T of N classical harmonic oscillators, for which the energy of one oscillator is $E = p^2/(2m) + Kx^2/2$ (p is the momentum, m the mass and K the spring constant). Compare the results with those of the quantum oscillators: in which temperature ranges are the results similar and in which ranges are they different?

Exercise 5.3

A linear harmonic oscillator has a magnetic moment dependent on its position. The energy of one oscillator is $E = p^2/(2m)+Kx^2/2-\gamma Hx$ (p is the momentum, m the mass, K the spring constant, H the magnetic field and γ is a constant).

(a) Calculate the one-oscillator partition function in classical mechanics.

   Hint: Use the identity $Ax^2 - Bx = A(x - B/(2A))^2 - B^2/(4A)$.

(b) Calculate the energy of N identical oscillators as a function of T and H.
(c) Calculate the magnetization of N oscillators.

   Hint: $\int_{-\infty}^{\infty} \exp(-x^2)\, dx = \sqrt{\pi}/2$.

Exercise 5.4

One considers again N linear classical harmonic oscillators with a constant magnetic moment b. The magnetic field H is the $+x$ direction. The magnetic energy is equal to $-bH$ if $x > 0$ and is equal to bH if $x < 0$.

(a) Calculate the one-oscillator partition function in classical mechanics.
(b) Calculate the energy of N identical oscillators as a function of T and H.
(c) Calculate the magnetization of N oscillators.

   Hint: $\int_{-\infty}^{\infty} \exp(-x^2)\, dx = \sqrt{\pi}/2$.

Exercise 5.5

Indistinguishable particles have two possible energies $E_1 = 0$ and $E_2 = e$, and they are in contact with a reservoir at temperature T. The walls of the reservoir are permeable to particles.

(a) Give an expression of the grand partition function as an infinite series, in the classical case.

(b) At what condition the series will converge? Express the condition with the help of the chemical potential, T and e.

(c) Suppose that the series of the grand partition function converges. Calculate N, the mean number of particles with the help of the chemical potential, T and e.

(d) Find the chemical potential as a function of N, T and e. Calculate the energy as function of N, T and e.

(e) Check if the condition of convergence found in (b) is verified.

Chapter 6

Exercise 6.1

(a) Calculate the thermal properties $(E(T), S(T)$ and $C(T))$ $(C(T)$ is the specific heat at constant area) of a gas of photons located on a surface of area A. Give the emission spectrum $K(\lambda)$.

(b) Same problem for a gas of photons in $1D$, on a line of length L.

Exercise 6.2

Give the expression of the emission spectrum of a photon gas in a volume V as a function of the frequency ω. Give the limits of the spectrum for $\hbar\omega/(k_B T) \ll 1$ and $\hbar\omega/(k_B T) \gg 1$.

Exercise 6.3

Calculate the emission spectrum of a gas of fermions without a fixed number (volume V, temperature T). Express the spectrum as a function of the wavelength and as a function of the frequency. Give the limits of the spectrum for $\hbar\omega/(k_B T) \ll 1$ and for $\hbar\omega/(k_B T) \gg 1$. Compare with the case of photons.

Exercise 6.4

Particles without a fixed number can have only two possible energies: ε_1 and $\varepsilon_2 > \varepsilon_1$. The two levels have the same degeneracy G (G states for each energy).

(a) The particles are bosons. Calculate the energy $E(T)$, the specific heat $C(T)$ and the mean number of particles $N_1(T)$ and $N_2(T)$ in the two levels. Give for these four quantities the limits for $T \to 0$ and $T \to \infty$.

(b) The same questions, if the particles are fermions.

(c) Plot the specific heats of bosons and of fermions on the same graph.

(d) Plot the ratios N_2/N_1 as a function of T for the bosons and for the fermions on the same graph.

Hint: The graphs are only qualitative.

Chapter 7

Exercise 7.1

N atoms are located on a surface forming a $2D$ solid of area A. Calculate the energy $E(T)$ following the Debye model. Give the limits at low and high temperatures of $E(T)$ and of the specific heat $C_v(T)$.

Exercise 7.2

Give an expression of the mean number of phonons for a $3D$ solid as a function of the temperature, and find the limits for $T \to 0$ and for $T \to \infty$.

Exercise 7.3

Give an expression of the mean number of phonons for a $2D$ solid as a function of the temperature, and find the limits for $T \to 0$ and for $T \to \infty$.

Exercise 7.4

Verify the relationship $PV = KE$ for a phonon gas (E is the energy, P is the pressure and V is the volume) in the limits of low and high temperatures.

Hint: Find the expression of the free energy F, and its limits. See Exercise 4.4.

Chapter 8

Exercise 8.1
A gas of N bosons with mass m is located onto a surface with area A.

(a) Write the equation from which it is possible to calculate the Bose–Einstein condensation temperature if it takes place. Show that the Bose–Einstein condensation does not take place.

   Hint: You have two possibilities to answer. One is to calculate the integral, knowing that $\int dx [\exp(x) - 1]^{-1} = \mathrm{Ln}[1 - \exp(-x)]$. In the second, you do not need to calculate the integral. Try both methods.

(b) Calculate the chemical potential μ as a function of T and find the limits for low and high temperatures.

   Hint: $\int dx [C \exp(x) - 1]^{-1} = \mathrm{Ln}[1 - C^{-1}\exp(-x)] (C > 1)$.

(c) Show that $d\mu/dT$ goes to zero with T. Sketch a qualitative graph of $\mu(T)$.

(d) Show that the ratio $\mu/(k_B T)$ goes to zero with T.

(e) Try to calculate the energy at low temperatures and show that the specific heat at constant volume varies linearly with T.

   Hint: Use an approximation based on the question. $\int_0^\infty x [\exp(x) - 1]^{-1} = \pi/12$.

Exercise 8.2
The dimension of the space can be only 3, 2 or 1. However, suppose that there are spaces with a non-integer dimension d. From the results of problem 4.2 one can write the momentum density of state as

$$g(p)\, dp = A p^{d-1}\, dp,$$

where d is the dimension of the space. Find what dimensions between 3 and 2 are for which one can observe a Bose–Einstein condensation.

Hint: You do not need to calculate any integral to answer this question.

Exercise 8.3

N bosons have the following properties:

1. They are located in a volume V.
2. The energy E of a boson is equal to pc (p is the momentum, c the velocity of light).

(a) Calculate the Bose–Einstein condensation temperature.
(b) Calculate the ratio N_1/N (N_1 is the number of bosons in the lowest level) as a function of the temperature in the condensed regime.
(c) Calculate the energy $E(T)$ in the condensed regime.

Hint: $\int_0^\infty x^2[\exp(x) - 1]^{-1} \approx 2.404$; $\int_0^\infty x^3[\exp(x) - 1]^{-1} \approx 6.5$.

Chapter 9

Exercise 9.1

(a) Calculate the entropy of a gas of fermions (mass m) in a volume V at low temperature ($T \neq 0$).
(b) Calculate the pressure of this gas at $T = 0$ and $T \neq 0$ (but not too high).

Exercise 9.2

For N electrons in a volume V at temperature T, the energy E is related to the linear momentum p as $E = pc$ (c is the light velocity).

(a) Calculate the Fermi level and the energy for $T = 0$.
(b) Show that for $T \neq 0$ (but not too high) the specific heat at constant volume is linear in T.

Exercise 9.3

The chemical potential, or the Fermi level, of a gas of electrons is positive at $T = 0$.

(a) Show that in a process at constant entropy and constant volume, the energy is increased if the number of particles in this gas is increased.
(b) Try to give a physical explication for this result, without calculation.

Exercise 9.4

The energy spectrum of N electrons in a volume V is as follows:

1. There is one level (energy $E = -\varepsilon$) with degeneracy N (one state for one electron).
2. For energy equal or larger than 0 there is a continuum of energy up to infinity.

(a) Find the position of the Fermi level at $T = 0$ (without calculation).
(b) Give an expression for the number n_1 of free electrons (as a function of T, N, ε and the Fermi level μ).
(c) Give an expression for the number of electrons n_2 on the level with energy $-\epsilon$ (as a function of T, N, ε and the Fermi level μ).
(d) Give an expression for the missing places $N - n_2$ on the level with energy $-\epsilon$.
(e) Find the Fermi level as a function of T (for low T) and calculate n_1 as a function of T.

Hint: For the last question, find the approximate expression for $N - n_2$, taking into account the position of the Fermi level.

Exercise 9.5

A gas of electrons is located on a surface of area A. (Electron gas in two dimensions).

(a) Calculate the Fermi level μ as a function of the temperature T. In what temperature does it becomes null? What is its value at $T = 0$?

Hint: $\int dx [C \exp(x) + 1]^{-1} = -\text{Ln}[1 + C^{-1} \exp(-x)]$.

(b) Calculate the derivative $d\mu/dT$ as a function of T and give its value for $T = 0$. Sketch a qualitative graph of $\mu(T)$.

(c) Show that for $T \neq 0$ (but not too high) C_V, the specific heat at constant volume, is linear with T.

(d) For a $2D$ boson gas at low temperatures the constant volume specific heat is also linear in T. Can you find an explanation why does C_V have the same temperature dependence for bosons and fermions?

Index